T0269463

SpringerBriefs in Physics

Editorial Board

Egor Babaev, University of Massachusetts, USA
Malcolm Bremer, University of Bristol, UK
Xavier Calmet, University of Sussex, UK
Francesca Di Lodovico, Queen Mary University of London, UK
Maarten Hoogerland, University of Auckland, New Zealand
Eric Le Ru, Victoria University of Wellington, New Zealand
Hans-Joachim Lewerenz, California Institute of Technology, USA
James Overduin, Towson University, USA
Vesselin Petkov, Concordia University, Canada
Charles H.-T. Wang, University of Aberdeen, UK
Andrew Whitaker, Queen's University Belfast, UK

More information about this series at http://www.springer.com/series/8902

Shinichiro Seki • Masahito Mochizuki

Skyrmions in Magnetic Materials

 Springer

Shinichiro Seki
Center for Emergent Matter Science (CEMS)
RIKEN
Wako, Japan

Masahito Mochizuki
Department of Physics and Mathematics
Aoyama Gakuin University
Sagamihara, Japan

ISSN 2191-5423   ISSN 2191-5431   (electronic)
SpringerBriefs in Physics
ISBN 978-3-319-24649-9   ISBN 978-3-319-24651-2   (eBook)
DOI 10.1007/978-3-319-24651-2

Library of Congress Control Number: 2015953776

Springer Cham Heidelberg New York Dordrecht London
© Springer International Publishing Switzerland 2016
This work is subject to copyright. All rights are reserved by the Publisher, whether the whole or part of the material is concerned, specifically the rights of translation, reprinting, reuse of illustrations, recitation, broadcasting, reproduction on microfilms or in any other physical way, and transmission or information storage and retrieval, electronic adaptation, computer software, or by similar or dissimilar methodology now known or hereafter developed.
The use of general descriptive names, registered names, trademarks, service marks, etc. in this publication does not imply, even in the absence of a specific statement, that such names are exempt from the relevant protective laws and regulations and therefore free for general use.
The publisher, the authors and the editors are safe to assume that the advice and information in this book are believed to be true and accurate at the date of publication. Neither the publisher nor the authors or the editors give a warranty, express or implied, with respect to the material contained herein or for any errors or omissions that may have been made.

Printed on acid-free paper

Springer International Publishing AG Switzerland is part of Springer Science+Business Media (www.springer.com)

Contents

Chapter 1
Theoretical Model of Magnetic Skyrmions

Abstract Skyrmions were originally proposed by Tony Skyrme in the 1960s to account for the stability of hadrons in particle physics as a topological solution of the non-linear sigma model. Bogdanov and his collaborators theoretically predicted their realisation in chiral-lattice ferromagnets with finite Dzyaloshinskii–Moriya interaction due to the lack of spatial inversion symmetry. In this chapter, an overview of theoretical aspects of magnetic skyrmions is provided.

1.1 What Is a Skyrmion?

Keen competition among interactions in magnets often gives rise to non-collinear or non-coplanar spin structures such as vortices, domain walls, bubbles and spirals. These spin structures endow hosting materials with interesting physical properties and useful device functions, which have attracted intense research interest from viewpoints of fundamental science and technical applications. For example, domain walls and vortices in metallic ferromagnets can be driven by spin-polarised electric currents [1–3], and their application to magnetic storage devices such as race-track memory is anticipated [4]. Magnetic spirals in insulating magnets often exhibit rich magnetoelectric cross-correlation phenomena due to the coupling between magnetism and electricity through the generation of ferroelectric polarisation via a relativistic spin–orbit interaction [5–7]. In addition to these spin structures, magnetic skyrmions, vortex-like swirling spin structures characterised by a quantised topological number, are attracting considerable research attention because it has turned out that their peculiar response dynamics to external fields hold highly promising properties with applications to spintronic device functions [8–10].

Skyrmions were originally proposed by Tony Skyrme in the 1960s to account for the stability of hadrons as quantised topological defects in the three-dimensional (3D) non-linear sigma model [11, 12]. They have now turned out to be highly relevant to a spin structure in condensed-matter systems. A magnetic skyrmion comprises spins pointing in all directions wrapping a sphere similar to a hedgehog, as shown in Fig. 1.1a. The number of such wrappings corresponds to a topological invariant, and thus, the skyrmion has topologically protected stability. It has been found that skyrmions are indeed realised in quantum Hall ferromagnets [13, 14],

© Springer International Publishing Switzerland 2016

S. Seki, M. Mochizuki, *Skyrmions in Magnetic Materials*, SpringerBriefs in Physics, DOI 10.1007/978-3-319-24651-2_1

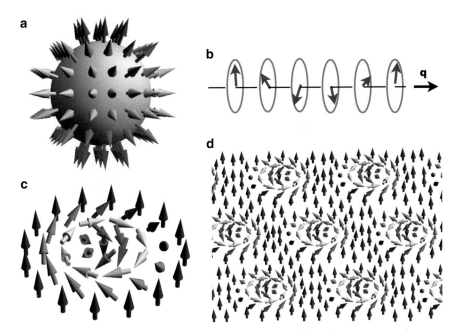

Fig. 1.1 (**a**) Schematic of the original hedgehog-type skyrmion proposed by Tony Skyrme in the 1960s, whose magnetisations point in all directions wrapping a sphere. (**b**) Schematic of the helical state realised in chiral-lattice magnets as a consequence of the competition between the Dzyaloshinskii–Moriya and ferromagnetic exchange interactions. (**c**) Schematic of a skyrmion recently discovered in chiral-lattice magnets, which corresponds to a projection of the hedgehog-type skyrmion on a two-dimensional (2D) plane. Its magnetisations also point in all directions wrapping a sphere. (**d**) Schematic of the skyrmion crystal realised in chiral-lattice magnets under an external magnetic field in which skyrmions are hexagonally packed to form a triangular lattice

ferromagnetic monolayers [15], doped layered antiferromagnets [16], liquid crystals [17] and Bose–Einstein condensates [18]. Recently, the realisation of magnetic skyrmions in chiral-lattice magnets was theoretically predicted [19–21] and later experimentally discovered [6, 22].

1.2 Stabilisation of Magnetic Skyrmions

There are several mechanisms for skyrmion formation in magnets. One major mechanism is the competition between the Dzyaloshinskii–Moriya and ferromagnetic exchange interactions [19–21]. In chiral-lattice ferromagnets without spatial inversion symmetry, such as B20 compounds (MnSi, FeGe, $Fe_{1-x}Co_xSi$) and copper oxoselenite Cu_2OSeO_3, the Dzyaloshinskii–Moriya interaction, which originates from the relativistic spin–orbit coupling, becomes finite [23, 24]. In a continuum spin model, the Dzyaloshinskii–Moriya interaction is expressed by

$$\mathscr{H}_{\mathrm{DM}} \propto \int \mathrm{d}\mathbf{r}\mathbf{M} \cdot (\nabla \times \mathbf{M}), \tag{1.1}$$

where $\mathbf{M}$ is the classical magnetisation vector. This interaction alone favours a rotating magnetisation alignment with a turn angle of 90° and competes with the ferromagnetic exchange interaction that favours a collinear ferromagnetic spin alignment. As a result of their competition, a helical spin order with a uniform turn angle shown in Fig. 1.1b is realised in the absence of an external magnetic field [25–28]. On the application of a weak magnetic field, skyrmions appear as vortex-like topological spin textures shown in Fig. 1.1c in a plane normal to the field irrespective of field direction. In a skyrmion, magnetisations are parallel to an applied magnetic field at its periphery but antiparallel at its centre. This spin structure corresponds to a projection of the original hedgehog-type skyrmion on a 2D plane. The topological nature of this projected skyrmion is characterised by the topological invariant

$$\mathscr{G} = \int \mathrm{d}^2 r \left(\frac{\partial \hat{\mathbf{n}}}{\partial x} \times \frac{\partial \hat{\mathbf{n}}}{\partial y} \right) \cdot \hat{\mathbf{n}}, \tag{1.2}$$

where $\hat{\mathbf{n}} = \mathbf{M}/|\mathbf{M}|$ is the unit vector pointing in the direction of magnetisation. This quantity is a sum of solid angles spanned by three neighbouring magnetisations; for a single skyrmion, its value is given by $4\pi Q$, with $Q(= \pm 1)$ being the skyrmion number. The sign of Q corresponds to that of magnetisation at the skyrmion core, i.e. $Q = +1$ ($Q = -1$) for up (down) magnetisation at the core. Skyrmions often form a hexagonal lattice, the so-called skyrmion crystal shown in Fig. 1.1d. Magnetisations align ferromagnetically in a stacking direction to form rod-like or tube-like structures. Typically, skyrmions in chiral-lattice ferromagnets are 3–100 nm in size, which is determined by the ratio of the Dzyaloshinskii–Moriya interaction D to the ferromagnetic exchange interaction J.

Another major mechanism of skyrmion formation is the competition between magnetic dipole interaction and easy-axis anisotropy [29–32]. In thin-film specimens of ferromagnets with perpendicular easy-axis anisotropy, the anisotropy favours out-of-plane magnetisations, whereas a long-range magnetic dipole interaction favours in-plane magnetisations. Their competition results in a periodic stripe with spins rotating in a thin-film plane. An application of a magnetic field normal to the thin-film plane turns the stripe into a periodic arrangement of magnetic bubbles or skyrmions. Skyrmions or bubbles of this origin tend to be large, typically 3–100 μm in size, which is orders of magnitude larger than skyrmions in chiral-lattice ferromagnets. In addition to these two mechanisms, frustrated exchange interactions [33] and four-ring exchange interactions [15] have been theoretically proposed as origins of skyrmion formation. Skyrmions of these origins tend to be atomically small.

1.3 Model and Phase Diagrams

To describe the magnetism in MnSi as a prototypical chiral-lattice ferromagnet, the continuum spin model that was proposed by Bak and Jensen in 1980 [34] is as follows:

$$\mathscr{H} = \int d^3r \left[\frac{J}{2a}(\nabla \mathbf{M})^2 + \frac{D}{a^2}\mathbf{M} \cdot (\nabla \times \mathbf{M}) \right.$$

$$- \frac{1}{a^3}\mathbf{B} \cdot \mathbf{M}$$

$$+ \frac{A_1}{a^3}(M_x^4 + M_y^4 + M_z^4)$$

$$\left. - \frac{A_2}{2a}[(\nabla_x M_x)^2 + (\nabla_y M_y)^2 + (\nabla_z M_z)^2] \right]. \qquad (1.3)$$

The first and second terms represent the ferromagnetic exchange interaction ($J > 0$) and the Dzyaloshinskii–Moriya interaction, respectively. The third term denotes the Zeeman coupling to an external magnetic field $\mathbf{B}$. The fourth and fifth terms are magnetic anisotropies allowed by a cubic crystal symmetry, but they turn out to play a minor role as far as realistically small values of A_1 and A_2 are considered. Here, a is the lattice constant.

In Ref. [35], the stability of a skyrmion-crystal phase was theoretically studied based on this model by writing the Ginzburg–Landau free energy functional near T_c as

$$F[\mathbf{M}] = \int d^3r \left[r_0 \mathbf{M}^2 + J(\nabla \mathbf{M})^2 + 2D\mathbf{M} \cdot (\nabla \times \mathbf{M}) \right.$$

$$\left. + U\mathbf{M}^4 - \mathbf{B} \cdot \mathbf{M} \right]. \qquad (1.4)$$

When a uniform component of magnetisation $\mathbf{M}_{uniform}$ is induced by a magnetic field, we obtain the term

$$\sum_{\mathbf{q}_1, \mathbf{q}_2, \mathbf{q}_3} (\mathbf{M}_{uniform} \cdot \mathbf{m}_{\mathbf{q}_1})(\mathbf{m}_{\mathbf{q}_2} \cdot \mathbf{m}_{\mathbf{q}_3})\delta(\mathbf{q}_1 + \mathbf{q}_2 + \mathbf{q}_3) \qquad (1.5)$$

from the quartic term in Eq. (1.4), where $\mathbf{m}_{\mathbf{q}}$ is the Fourier component of $\mathbf{M}(\mathbf{r})$. Wave vectors $\mathbf{q}_1$, $\mathbf{q}_2$ and $\mathbf{q}_3$ should have a fixed modulus determined by two competing gradient terms, i.e. the ferromagnetic exchange term and Dzyaloshinskii–Moriya term. In addition, the energy change should be proportional to $\mathbf{M}_{uniform} \cdot \hat{n}$ by symmetry, where $\hat{n}$ is a vector normal to the plane spanned by the three wave vectors. Therefore, one can gain energy from this term for the skyrmion-crystal structure characterised as a superposition of three helices with equal pitch length, equal chirality and a relative angle of 120° propagating in the plane normal to a magnetic field. Magnetisation in this structure is expressed by

$$\mathbf{M}(\mathbf{r}) \approx \mathbf{M}_{\text{uniform}} + \sum_{i=1}^{3} \mathbf{M}_{\mathbf{Q}_i}(\mathbf{r} + \Delta\mathbf{r}_i) \tag{1.6}$$

where

$$\mathbf{M}_{\mathbf{Q}_i}(\mathbf{r}) = A\left[\mathbf{n}_{i1}\cos(\mathbf{Q}_i \cdot \mathbf{r}) + \mathbf{n}_{i2}\sin(\mathbf{Q}_i \cdot \mathbf{r})\right] \tag{1.7}$$

is the magnetisation of a single helix with amplitude A and $\mathbf{Q}_i \cdot \Delta\mathbf{r}_i$ is its phase. Two unit vectors, $\mathbf{n}_{i1}$ and $\mathbf{n}_{i2}$, are orthogonal to each other as well as to $\mathbf{Q}_i$. The three wave vectors, $\mathbf{Q}_i$ ($i = 1, 2, 3$), satisfy the relation

$$\sum_{i=1}^{3} \mathbf{Q}_i = \mathbf{0}. \tag{1.8}$$

It is revealed that within the mean-field analysis of the Ginzburg–Landau theory, the skyrmion-crystal phase is always higher in energy than the conical spin phase with a propagation wave vector parallel to an applied magnetic field. The free energy is given by

$$\exp(-G) = \int D\mathbf{M}e^{-F[\mathbf{M}]}, \tag{1.9}$$

and $G(\mathbf{B})$ is equal to $\min F[\mathbf{M}]$ within the mean-field approximation. It is found that the appropriate treatment of thermal fluctuations beyond the mean-field approximation is necessary to reproduce the skyrmion-crystal phase. Indeed, the incorporation of the Gaussian thermal fluctuation turns out to reproduce the skyrmion-crystal phase in a narrow window of temperature and magnetic field on the verge of a paramagnetic-conical phase boundary. Here, the free energy is given by

$$G \approx F[\mathbf{M}_0] + \frac{1}{2}\log\left[\det\left(\frac{\delta^2 F}{\delta\mathbf{M}\delta\mathbf{M}}\right)_{\mathbf{M}=\mathbf{M}_0}\right] \tag{1.10}$$

where $\mathbf{M}_0$ is the mean-field spin configuration.

Although the skyrmion-crystal phase is rather unstable in a 3D model and in bulk specimens, it turns out to have greater stability in a 2D system [36]. The continuum model in two dimensions is given by

$$\mathcal{H} = \int d^2r \left[\frac{J}{2} (\nabla \mathbf{M})^2 + \frac{D}{a} \mathbf{M} \cdot (\nabla \times \mathbf{M}) \right.$$

$$- \frac{1}{a^2} \mathbf{B} \cdot \mathbf{M}$$

$$+ \frac{A_1}{a^2} (M_x^4 + M_y^4 + M_z^4)$$

$$\left. - \frac{A_2}{2} [(\nabla_x M_x)^2 + (\nabla_y M_y)^2] \right]. \tag{1.11}$$

Starting from this continuum model, we obtain a lattice spin model (i.e. a classical Heisenberg model on a square lattice) by dividing the space into square meshes:

$$\mathcal{H} = -J \sum_i \mathbf{m}_i \cdot (\mathbf{m}_{i+\hat{x}} + \mathbf{m}_{i+\hat{y}})$$

$$- D \sum_i (\mathbf{m}_i \times \mathbf{m}_{i+\hat{x}} \cdot \hat{x} + \mathbf{m}_i \times \mathbf{m}_{i+\hat{y}} \cdot \hat{y})$$

$$- \mathbf{B} \cdot \sum_i \mathbf{m}_i$$

$$+ A_1 \sum_i [(m_i^x)^4 + (m_i^y)^4 + (m_i^z)^4]$$

$$- A_2 \sum_i (m_i^x m_{i+\hat{x}}^x + m_i^y m_{i+\hat{y}}^y), \tag{1.12}$$

where $\hat{x}$ and $\hat{y}$ are bond vectors on the square lattice. Here, $\mathbf{m}_i$ is a classical vector with a constant norm of $|\mathbf{m}_i| = m$, which represents magnetisation at the ith lattice site occupying an area of a^2. As far as magnetic structures with long spatial modulation, such as helices and skyrmions, are considered, influence from a background lattice structure becomes negligible, which justifies the simple meshing of space into squares without considering real complicated crystal structures.

For slowly varying magnetic orders, we can regard an assembly of magnetisations in a larger area of $(xa)^2$ as a coarse-grained magnetic unit. When one represents this magnetic unit as $\mathbf{m}_i$ with a norm of $|\mathbf{m}_i| = m$, model parameters (J, D, B_z, A_1, A_2) given by Eq. (1.12) should be rescaled as (J, xD, $x^2 B_z$, $x^2 A_1$, A_2) so as to make the energy scale unchanged. A Monte-Carlo analysis of this lattice spin model reveals a rich phase diagram at $T = 0$, as shown in Fig. 1.2, which includes not only a hexagonal skyrmion-crystal phase but also various phases of unusual magnetic patterns [36].

Figure 1.3 displays a phase diagram in plane of temperature and magnetic field (a T-B phase diagram) obtained by Monte-Carlo calculations for this 2D lattice spin model without anisotropy terms ($A_1 = 0$ and $A_2 = 0$) [22]. We find that the skyrmion-crystal phase occupies a wide region in the phase diagram and occurs

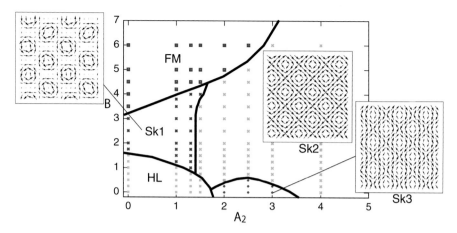

Fig. 1.2 Theoretical phase diagram in the plane of the anisotropy A_2 and the magnetic field B at $T = 0$ for the 2D classical Heisenberg model given by Eq. (1.12) with $\mathbf{B} = (0, 0, B)$ obtained by Monte-Carlo calculations. The calculations are performed for a system with 18×18 sites with a periodic boundary condition. Parameters are set to be $J = 1$, $D = \sqrt{6}$, $A_1 = 0.5$ and $m = 1$. *Sk1* is a triangular skyrmion-crystal phase, whereas *Sk2* and *Sk3* are different types of skyrmion-crystal phases. *HL* and *FM* denote helical and field-polarised ferromagnetic phases, respectively

even down to very low temperatures. Moreover, it turns out that the skyrmion-crystal phase emerges even at $T = 0$. The zero-temperature phase diagram as a function of magnetic field B was obtained by a combination of Monte-Carlo technique and Landau–Lifshitz–Gilbert simulation (see Fig. 1.4), in which the skyrmion-crystal phase emerges in the range of moderate field strength sandwiched between helical and ferromagnetic phases. In the absence of a magnetic field, a helical order with a propagating wave vector $\mathbf{Q}$ confined within the plane occurs. Spins in this state are rotating in a plane perpendicular to $\mathbf{Q}$ to form a proper screw structure. As B_z increases, this helical state turns into a triangular skyrmion-crystal state. As B_z further increases, a field-polarised ferromagnetic order, in which all spins point parallel to $\mathbf{B}$, occurs. This phase diagram reproduces experimental phase diagrams for thin-film samples of MnSi and Cu_2OSeO_3. We can indeed find that these three phases successively appear in the experimental B-T phase diagram along a vertical line at the lowest temperature.

The enhanced stability of the skyrmion-crystal phase in the 2D system can be understood as follows: when the magnetic field $\mathbf{B}$ is applied to a 3D system, the conical order with propagating wave vector $\mathbf{Q} \parallel \mathbf{B}$ and net magnetisation $\mathbf{M} \parallel \mathbf{B}$ becomes stabilised. However, spins can no longer rotate to form a conical order in the 2D system when $\mathbf{B}$ is applied normal to the plane. As a result, the skyrmion-crystal phase attains relative stability against the conical state. Such a destabilisation of the conical state occurs not only in a pure 2D system but also in thin-film systems with thicknesses less than the helical wavelength. This prediction is confirmed by the real-space observation of a triangular skyrmion crystal in $Fe_{0.5}Co_{0.5}Si$ specimens

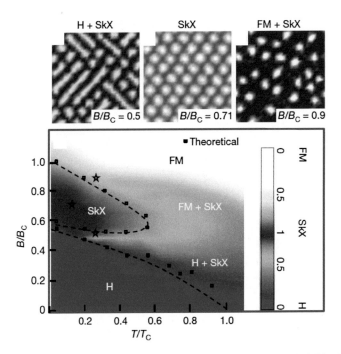

Fig. 1.3 Theoretical phase diagram in the plane of temperature and magnetic field B for the 2D classical Heisenberg model given by Eq. (1.12) with $A_1 = 0$, $A_2 = 0$ and $\mathbf{B} = (0, 0, B)$ obtained by Monte-Carlo calculations, in which the phase change of a spin texture is represented by the contour mapping of skyrmion density. *H*, helical structure; *SkX*, skyrmion-crystal structure; *FM*, ferromagnetic structure. *FM+SkX* and *H+SkX* denote coexistences of the two magnetic structures (Reproduced from Ref. [22])

of conventional film thickness (several tens of nanometres) using the Lorentz force microscopy in a wide temperature and magnetic field range [22].

Modulation periods of helical and skyrmion-crystal structures are determined by the ratio D/J, i.e. the Dzyaloshinskii–Moriya interaction to ferromagnetic exchange interaction. Writing the energy as a function of the turn angle θ in the helical structure, we obtain the relation from the saddle-point equation as follows:

$$\frac{D}{\sqrt{2}J} = \tan \theta = \tan(2\pi/\lambda_m), \qquad (1.13)$$

where λ_m is the pitch length of the helix in units of the lattice constant. For example, λ_m takes $\sim$100 sites for $|D/J| = 0.09$ and $\sim$35 sites for $|D/J| = 0.27$; if one assumes a typical lattice constant of 5 Å, these pitch lengths correspond to 50 and 17.5 nm, respectively, which reproduce those in Cu_2OSeO_3 ($\lambda_m = 50$ nm) and MnSi ($\lambda_m = 18$ nm). On the other hand, the period of the skyrmion crystal is approximately $2/\sqrt{3}$ times as long as that of the helix.

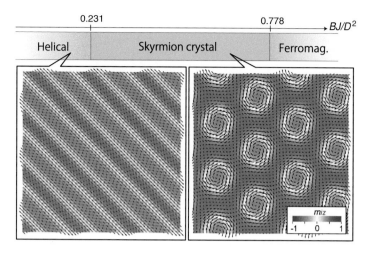

Fig. 1.4 Theoretical phase diagram as a function of magnetic field B at $T = 0$ for the 2D classical Heisenberg model given by Eq. (1.12) with $A_1 = 0$, $A_2 = 0$ and $\mathbf{B} = (0, 0, B)$ obtained by a combination of Monte-Carlo technique and Landau–Lifshitz–Gilbert simulation. The hexagonal skyrmion-crystal phase emerges in a moderate field strength region sandwiched between helical and field-polarised ferromagnetic phases (Reproduced from Ref. [37])

A lattice spin model in three dimensions is also studied. After dividing the space into cubic meshes, we obtain a classical Heisenberg model on the cubic lattice from the continuum model:

$$\mathcal{H} = -J \sum_i \mathbf{m}_i \cdot (\mathbf{m}_{i+\hat{x}} + \mathbf{m}_{i+\hat{y}} + \mathbf{m}_{i+\hat{z}})$$

$$-D \sum_i (\mathbf{m}_i \times \mathbf{m}_{i+\hat{x}} \cdot \hat{x} + \mathbf{m}_i \times \mathbf{m}_{i+\hat{y}} \cdot \hat{y} + \mathbf{m}_i \times \mathbf{m}_{i+\hat{z}} \cdot \hat{z})$$

$$-\mathbf{B} \cdot \sum_i \mathbf{m}_i$$

$$+A_1 \sum_i [(m_i^x)^4 + (m_i^y)^4 + (m_i^z)^4]$$

$$-A_2 \sum_i (m_i^x m_{i+\hat{x}}^x + m_i^y m_{i+\hat{y}}^y + m_i^z m_{i+\hat{z}}^z). \tag{1.14}$$

In the case of three dimensions, model parameters (J, D, B_z, A_1 and A_2) Eq. (1.14) should be rescaled by (xJ, x^2D, x^3B_z, x^3A_1 and xA_2) when one represents an assembly of magnetisations in a larger volume of $(xa)^3$ with coarse-grained magnetisation $\mathbf{m}_i$. This 3D lattice model with $A_1 = 0$ and $A_2 = 0$ is analysed using Monte-Carlo technique to examine the phase diagram in the temperature–magnetic field plane [38]. The obtained phase diagram is displayed in the inset of Fig. 1.5. In

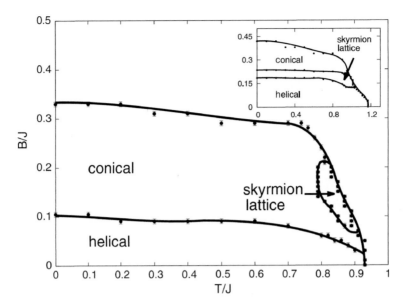

Fig. 1.5 Theoretical phase diagram in the temperature–magnetic field plane obtained by Monte-Carlo calculations for the 3D classical Heisenberg model, in which further neighbour interactions J' and D' are added to the model given in Eq. (1.14) so as to compensate for induced anisotropies due to spatial discretisation. The calculations are performed for a system with $30 \times 30 \times 30$ sites with a periodic boundary condition with $A_1 = 0$, $A_2 = 0$ and $\mathbf{B} = (0, 0, B)$. Parameters are set to be $J = 1$, $D = 0.727$, $J' = J/16$, $D' = D/8$ and $m = 1$. We find that the skyrmion-crystal phase exists only in a small window near a conical-paramagnetic phase diagram, which is in agreement with experiments. The *inset* shows a theoretical phase diagram for a 3D lattice spin model without anisotropy compensation. Conditions for the calculations are equivalent to those for the above case except that $J' = 0$ and $D' = 0$. We find that the skyrmion-crystal phase remains even down to $T \to 0$, which contradicts the experimental results (Reproduced from Ref. [38])

the simulations, the ratio D/J is fixed at $D/J \sim 0.727$, which gives a pitch length of ~ 10 lattice sites in the helical phase. The simulation is performed for a system with $N = 30 \times 30 \times 30$ sites, with a periodic boundary condition where nine skyrmion tubes are hexagonally packed into the obtained skyrmion-crystal phase. Apparently, there is a discrepancy between this theoretical phase diagram and the experimental ones for the bulk specimens of MnSi and Cu_2OSeO_3, as shown in Fig. 1.6a and c. Experimentally, the skyrmion-crystal phase in the bulk specimens appears only at finite temperatures on the verge of a paramagnetic-helical phase boundary. In turn, the skyrmion phase remains stable even for $T \to 0$ in the theoretical phase diagram shown in the inset of Fig. 1.5.

In Ref. [38], this discrepancy was attributed to the finite-size effect or anisotropies due to the discretisation of the continuum model. The ferromagnetic exchange term and Dzyaloshinskii–Moriya term in Eq. (1.14) on the cubic lattice after Fourier transformation read

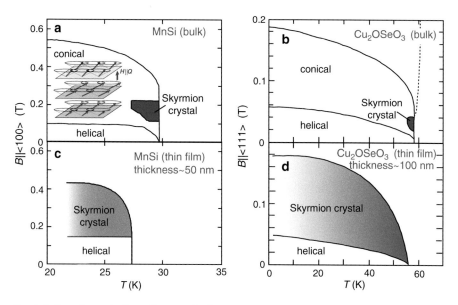

Fig. 1.6 Experimental phase diagrams in the temperature–magnetic field plane for (**a**) a bulk sample of MnSi [35], (**b**) a bulk sample of Cu_2OSeO_3 [40], (**c**) a thin-film sample of MnSi [39] and (**d**) a thin-film sample of Cu_2OSeO_3 [40]

$$\mathcal{H}_{FM} + \mathcal{H}_{DM} = \sum_{\mathbf{k}} \alpha_{\mathbf{k}} \mathbf{m}_{\mathbf{k}} \cdot \mathbf{m}_{-\mathbf{k}}$$

$$+ \sum_{\mathbf{k}} (\beta_{x\mathbf{k}}\hat{x} + \beta_{y\mathbf{k}}\hat{y} + \beta_{z\mathbf{k}}\hat{z}) \cdot (\mathbf{m}_{\mathbf{k}} \times \mathbf{m}_{-\mathbf{k}}), \qquad (1.15)$$

where

$$\alpha_{\mathbf{k}} = J \left(\cos(k_x a) + \cos(k_y a) + \cos(k_z a) \right)$$

$$= -3J + \frac{a^2 J}{2}(k_x^2 + k_y^2 + k_z^2)$$

$$- \frac{a^4 J}{24}(k_x^4 + k_y^4 + k_z^4) + \mathcal{O}(k^6), \qquad (1.16)$$

$$\beta_{\gamma\mathbf{k}} = -D\sin(k_y a) = -aDk_\gamma + \frac{a^3 D}{6}k_\gamma^3 + \mathcal{O}(k^5). \qquad (1.17)$$

These expressions contain terms of higher order in momentum. In contrast, the Fourier transform of the ferromagnetic exchange term in the continuum model contains only quadratic terms, whereas that of the Dzyaloshinskii–Moriya term contains only a linear term. For magnetic orders with finite wave vectors, contributions of higher-order terms are not negligible in a lattice model, and we need to add further neighbour interactions $\mathscr{H}'$ to the Hamiltonian so as to compensate for induced anisotropies and to achieve better approximation to the continuum model without breaking symmetries of an underlying system. The term $\mathscr{H}'$ is given by

$$\mathscr{H}' = J' \sum_i \mathbf{m}_i \cdot (\mathbf{m}_{i+2\hat{x}} + \mathbf{m}_{i+2\hat{y}} + \mathbf{m}_{i+2\hat{z}})$$

$$+ D' \sum_i (\mathbf{m}_i \times \mathbf{m}_{i+2\hat{x}} \cdot \hat{x} + \mathbf{m}_i \times \mathbf{m}_{i+2\hat{y}} \cdot \hat{y} + \mathbf{m}_i \times \mathbf{m}_{i+2\hat{z}} \cdot \hat{z}). \qquad (1.18)$$

Consequently, the full $\alpha_{\mathbf{k}}$ and $\beta_{\gamma \mathbf{k}}$ are respectively given by

$$\alpha_{\mathbf{k}} = -3(J - J') + \frac{a^2}{2}\left(J - 4J'\right)(k_x^2 + k_y^2 + k_z^2)$$

$$- \frac{a^4}{24}\left(J - 16J'\right)(k_x^4 + k_y^4 + k_z^4) + \mathscr{O}(k^6), \qquad (1.19)$$

$$\beta_{\gamma \mathbf{k}} = -a(D - 2D')k_\gamma + \frac{a^3}{6}(D - 8D')k_\gamma^3 + \mathscr{O}(k^5). \qquad (1.20)$$

These expressions indicate that anisotropies can be compensated by selecting $J' = J/16$ and $D' = D/8$. A Monte-Carlo simulation of this anisotropy-compensated lattice spin model successfully reproduces a phase diagram with a tiny skyrmion-crystal phase, as shown in Fig. 1.5, which is in excellent agreement with the experimental phase diagrams for bulk specimens.

References

1. J.C. Slonczewski, J. Magn. Magn. Mater. **159**, L1 (1996)
2. L. Berger, Phys. Rev. B **54**, 9353 (1996)
3. S.E. Barns, S. Maekawa, Phys. Rev. Lett. **95**, 107204 (2005)
4. S.S.P. Parkin, M. Hayashi, L. Thomas, Science **320**, 190 (2008)
5. Y. Tokura, Science **312**, 1481 (2006)
6. S.-W. Cheong, M. Mostovoy, Nat. Mater. **6**, 13 (2007)
7. H. Katsura, A.V. Balatsky, N. Nagaosa, Phys. Rev. Lett. **95**, 057205 (2005)
8. C. Pfleiderer, Nat. Phys. **7**, 673 (2011)
9. N. Nagaosa, Y. Tokura, Nat. Nanotech. **8**, 899 (2013)
10. A. Fert, V. Cros, J. Sampaio, Nat. Nanotech. **8**, 152 (2013)
11. T.H.R. Skyrme, Proc. R. Soc. A **260**, 127 (1961)
12. T.H.R. Skyrme, Nucl. Phys. **31**, 556 (1962)
13. S.L. Sondhi, A. Karlhede, S.A. Kivelson, E.H. Rezayi, Phys. Rev. B **47**, 16419 (1993)
14. M. Abolfath, J.J. Palacios, H.A. Fertig, S.M. Girvin, A.H. MacDonald, Phys. Rev. B **56**, 6795 (1997)
15. S. Heinze, K. von Bergmann, M. Menzel, J. Brede, A. Kubetzka, R. Wiesendanger, G. Bihlmayer, S. Blügel, Nat. Phys. **7**, 713 (2011)

16. I. Raičević, D. Popović, C. Panagopoulos, L. Benfatto, M.B. Silva Neto, E.S. Choi, T. Sasagawa, Phys. Rev. Lett. **106**, 227206 (2011)
17. D.C. Wright, N.D. Mermin, Rev. Mod. Phys. **61**, 385 (1989)
18. T.L. Ho, Phys. Rev. Lett. **81**, 742 (1998)
19. A.N. Bogdanov, D.A. Yablonskii, Sov. Phys. JETP **68**, 101 (1989)
20. A. Bogdanov, A. Hubert, J. Magn. Magn. Mat. **138**, 255 (1994)
21. U.K. Rößler, A.N. Bogdanov, C. Pfleiderer, Nature **442**, 797 (2006)
22. X.Z. Yu, Y. Onose, N. Kanazawa, J.H. Park, J.H. Han, Y. Matsui, N. Nagaosa, Y. Tokura, Nature **465**, 901 (2010)
23. I. Dzyaloshinskii, J. Phys. Chem. Solids **4**, 241 (1958)
24. T. Moriya, Phys. Rev. **120**, 91 (1960)
25. Y. Ishikawa, K. Tajima, D. Bloch, M. Roth, Solid State Commun. **19**, 525 (1976)
26. J. Beille, J. Voiron, M. Roth, Solid State Commun. **47**, 399 (1983)
27. B. Lebech, J. Bernhard, T. Freltoft, J. Phys. Condens. Matter **1**, 6105 (1989)
28. B. Lebech, P. Harrisa, J. Skov Pedersena, K. Mortensena, C.I. Gregoryb, N.R. Bernhoeftc, M. Jermyd, S.A. Browne, J. Magn. Magn. Mater. **140–144**, 119 (1995)
29. Y.S. Lin, J. Grundy, E.A. Giess, Phys. Lett. **23**, 485 (1973)
30. A.P. Malozemoff, J.C. Slonczewski, *Magnetic Domain Walls in Bubble Materials* (Academic, New York, 1979), pp. 306–314
31. T. Garel, S. Doniach, Phys. Rev. B **26**, 325 (1982)
32. T. Suzuki, J. Magn. Magn. Mater. **31–34**, 1009 (1983)
33. T. Okubo, S. Chung, H. Kawamura, Phys. Rev. Lett. **108**, 017206 (2012)
34. P. Bak, M.H. Jensen, J. Phys. C **13**, L881 (1980)
35. S. Mühlbauer, B. Binz, F. Jonietz, C. Pfleiderer, A. Rosch, A. Neubauer, R. Georgii, P. Böni, Science **323**, 915 (2009)
36. S.D. Yi, S. Onoda, N. Nagaosa, J.H. Han, Phys. Rev. B **80**, 054416 (2009)
37. M. Mochizuki, Phys. Rev. Lett. **108**, 017601 (2012)
38. S. Buhrandt, L. Fritz, Phys. Rev. B **88**, 195137 (2013)
39. A. Tonomura X.Z. Yu, K. Yanagisawa, T. Matsuda, Y. Onose, N. Kanazawa, H.S. Park, Y. Tokura, Nano Lett. **12**, 1673 (2012)
40. S. Seki, X.Z. Yu, S. Ishiwata, Y. Tokura, Science **336**, 198 (2012)

Chapter 2
Observation of Skyrmions in Magnetic Materials

Abstract Experimentally, skyrmion spin textures are observed in various magnetic systems with distinctive characteristics. In this chapter, some typical material environments, i.e. (1) non-centrosymmetric ferromagnets, (2) centrosymmetric ferromagnets with uniaxial anisotropy and (3) surface/interface of ferromagnetic monolayers, are introduced for realising skyrmion spin textures. (1) and (3) are systems with broken space-inversion symmetry, and thus, the Dzyaloshinskii–Moriya interaction is active and serves as a key source for stabilising skyrmion spin textures. For (2), in contrast, the breaking of space-inversion symmetry is not relevant, but an interplay between the dipole–dipole interaction and magnetic anisotropies is important for the realisation of magnetic skyrmions. In all cases, the typical size of a magnetic skyrmion ranges from sub-micrometre to nanometre, which implies that specific experimental techniques are required to identify the emergence of skyrmion spin textures directly.

2.1 Skyrmions in Non-centrosymmetric Magnets

We first introduce skyrmions in non-centrosymmetric ferromagnets [1–3]. In such environments with broken space-inversion symmetry, the Dzyaloshinskii–Moriya interaction favouring a canted spin arrangement becomes active [4, 5] and often stabilises a modulated spin texture. In this class of materials, all experimental reports of skyrmion formation have been made for chiral cubic ferro/ferrimagnets thus far. In Fig. 2.1, two material systems hosting skyrmion spin textures are indicated. One representative example is a series of metallic or semiconducting materials called B20 alloys (such as MnSi [6–10], $Fe_{1-x}Co_xSi$ [11–14], FeGe [15] and $Mn_{1-x}Fe_xGe$ [16, 17]), which have a common lattice form, as shown in Fig. 2.1a. They are usually characterised by an itinerant ferromagnetic exchange interaction. Another example is insulating Cu_2OSeO_3 [19–22]. This material contains two distinctive magnetic Cu^{2+} ($S = 1/2$) sites, as shown in Fig. 2.1b, and shows a local ferrimagnetic spin arrangement between them. These two systems belong to the same chiral cubic space group $P2_13$, indicating that the global symmetry or the overall nature of the Dzyaloshinskii–Moriya interaction is common between them. In both cases, their magnetic interactions consist of three hierarchical energy scales [23]. The strongest is the ferromagnetic or ferrimagnetic exchange interaction

© Springer International Publishing Switzerland 2016
S. Seki, M. Mochizuki, *Skyrmions in Magnetic Materials*, SpringerBriefs
in Physics, DOI 10.1007/978-3-319-24651-2_2

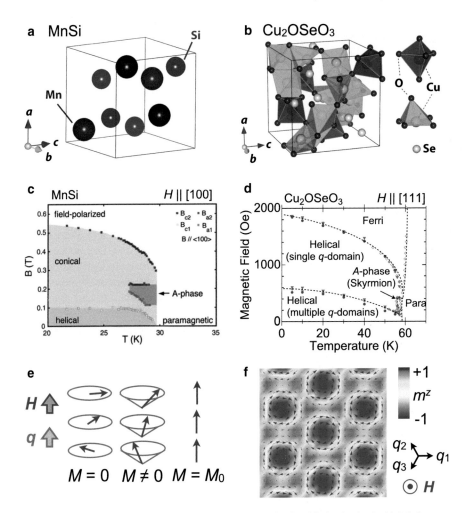

Fig. 2.1 (**a**), (**b**) Crystalline structures of (**a**) MnSi and (**b**) Cu_2OSeO_3, both of which belong to the chiral cubic space group $P2_13$. (**c**), (**d**) Temperature versus magnetic field phase diagrams for bulk samples of (**c**) MnSi and (**d**) Cu_2OSeO_3. Because of the isotropic nature of a cubic lattice, similar magnetic phase diagrams can be obtained for any direction of B. (**e**) Development of a helical spin texture characterised by a single magnetic modulation vector **q** as a function of an applied magnetic field B. (**f**) Schematic of a triangular lattice of skyrmions (i.e. skyrmion crystal) found for the above-mentioned materials, which is characterised by triple magnetic modulation vectors q_n ($n = 1, 2, 3$) within a plane normal to the B direction. Background colour indicates the out-of-plane component of local magnetisation m_z. Local magnetisation at the core (edge) of each skyrmion is anti-parallel (parallel) to the **B** direction (Reproduced from (**c**) Ref. [6] and (**d**)–(**f**) Ref. [19])

favouring a collinear spin arrangement, which is followed by the Dzyaloshinskii–Moriya interaction giving rise to a long-period modulation of spin texture. Magnetic anisotropy is relatively weak as compared with the above two interactions but still plays an important role in determining the spin modulation direction. Because the two systems are characterised by similar magnetic phase diagrams shown in Fig. 2.1c, d, we focus initially on the case of B20 alloys.

Figure 2.1c shows the phase diagram for a bulk sample of MnSi in the temperature (T)–magnetic field (B) plane [6]. Below the magnetic ordering temperature $T_c \sim 29$ K without an applied magnetic field, a helical spin order characterised by a single magnetic modulation vector $\mathbf{q}$ shown in Fig. 2.1e is stabilised as a result of the competition between the ferromagnetic exchange and Dzyaloshinskii–Moriya interactions. The favourable $\mathbf{q}$ direction is determined by magnetic anisotropy, and multiple equivalent helimagnetic $\mathbf{q}$ domains coexist because of the original cubic lattice symmetry. The period of the helical spin modulation $\lambda \sim 190$ Å is much larger than the crystallographic lattice constant $a \sim 4.56$ Å, implying a rather weak coupling of magnetic and atomic structures. The application of a magnetic field leads to a single-domain helical spin state so as to satisfy the relation $\mathbf{B} \parallel \mathbf{q}$ because anti-ferromagnetically aligned spins favour lying normal to an external $\mathbf{B}$. Here, a helical spin texture has a uniform magnetisation component along an applied $\mathbf{B}$ direction and thus can be considered to be conical. A further increase of the applied B realises a forced ferromagnetic state, as shown in Fig. 2.1e.

Another distinctive magnetic phase, the so-called 'A phase', has also been identified in a narrow B–T region just below T_c with a moderate magnitude of B, which causes anomalies in several macroscopic properties such as magnetic susceptibility, magnetoresistance [24], electronic spin resonance [25] and ultrasonic absorption [26]. While the spin texture in this A phase remained unresolved for several decades, Pfleiderer and co-authors revealed skyrmion lattice formation via small-angle neutron-scattering (SANS) experiments in 2009 [6]. In this method, an incident neutron beam was scattered by a sample and magnetic Bragg reflections within a reciprocal plane normal to the neutron-incident direction could be detected (see Fig. 2.2c). Figure 2.2a and b indicate the typical SANS data recorded for the magnetic A phase. Six-fold magnetic Bragg reflections always appear within a plane normal to B irrespective of the B direction. On the basis of these data, the spin texture $\mathbf{m}(\mathbf{r})$ described by the summation of three magnetic helices

$$\mathbf{m}(\mathbf{r}) \propto \mathbf{e}_z M_f + \sum_{a=1}^{3} [\mathbf{e}_z \cos(\mathbf{q}_a \cdot \mathbf{r} + \theta_a) + \mathbf{e}_a \sin(\mathbf{q}_a \cdot \mathbf{r} + \theta_a)], \qquad (2.1)$$

was proposed. $\mathbf{q}_a$ denotes one of the three magnetic modulation vectors normal to B, which forms an angle of $120°$ with respect to one another. $\mathbf{e}_z$ is a unit vector parallel to $\mathbf{B}$, and $\mathbf{e}_a$ is a unit vector orthogonal to both $\mathbf{e}_z$ and $\mathbf{q}_a$, defined such that all $\mathbf{q}_a \cdot (\mathbf{e}_z \times \mathbf{e}_a)$ have the same sign. M_f scales with the relative magnitude of net magnetisation along the B direction. When the phase shift of each of the magnetic helices θ_a takes some specific value, the spin texture of Eq. (2.1) can be considered to be a triangular lattice of skyrmions, i.e. the skyrmion crystal (SkX) shown in Fig. 2.1f. Combined with an additional theoretical calculation of free energy as well as the observation of the topological Hall effect [27], the formation of such

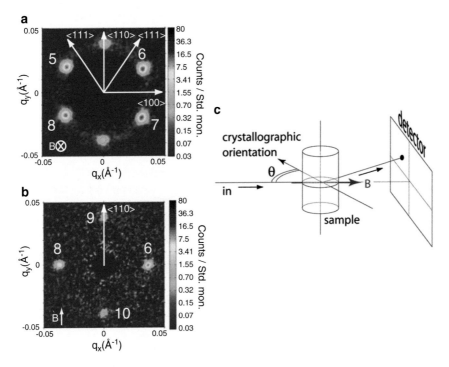

Fig. 2.2 Results of small-angle neutron-scattering (SANS) measurements performed at the magnetic A phase of bulk MnSi under a magnetic field (**a**) normal and (**b**) parallel to an observation plane. The background colour indicates the intensity of magnetic Bragg reflections at each point of reciprocal space. (**c**) Schematic of the experimental setup for the SANS experiment (Reproduced from Ref. [6])

a skyrmion crystal has been established for the A phase of bulk MnSi. Note that theories also predict that thermal fluctuation is essential for stabilising the skyrmion crystal state over a helical spin state in a chiral cubic ferromagnet (Sect. 1.3), which explains why a skyrmion crystal appears only in a limited temperature range just below T_c.

This breakthrough finding in the bulk sample is further followed by a real-space observation of a skyrmion crystal in thin-plate-shaped samples [12]. For this purpose, the Lorentz transmission electron microscopy (LTEM) technique is employed. When an electron beam passes through a magnetic material, each electron slightly changes its propagation direction as it experiences Lorentz force from local magnetisations in the sample. As a result, by taking over- and under-focused images and performing some additional numerical analysis (solving the magnetic transport-of-intensity equation (TIE)), we can obtain the real-space distribution of in-plane components of local magnetisations [28]. Figure 2.3a–f indicate the LTEM data taken for a $Fe_{1-x}Co_xSi$ ($x = 0.5$) single crystal with a thickness of several tens of nanometres [12]. Here, the thin-plate-shaped sample is prepared by thinning a

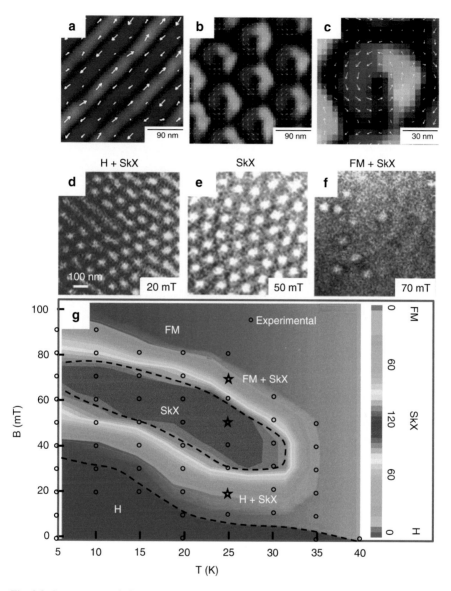

Fig. 2.3 Lorentz transmission electron microscopy (LTEM) data for $Fe_{0.5}Co_{0.5}Si$ with a thickness of several tens of nanometres. (**a**)–(**c**) Experimentally observed real-space images of lateral magnetisation distributions obtained through the transport-of-intensity equation (TIE) analysis of LTEM data: (**a**) Helical spin structure with $H = 0$, (**b**) skyrmion crystal structure for $H = 50$ mT applied normal to a thin plate and (**c**) a magnified view of (**b**). The *colour map* and *white arrows* represent magnetisation direction at each point. (**d**)–(**f**) Magnetic field dependence of LTEM (overfocus) images. (**g**) Temperature versus magnetic field phase diagram, where *H*, *SkX* and *FM* denote helical, skyrmion crystal and ferromagnetic spin states, respectively. The *colour bar* indicates skyrmion density per 10^{-12} m^2 (Reproduced from Ref. [12])

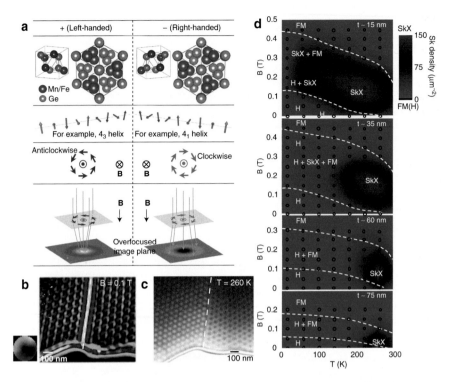

Fig. 2.4 LTEM data for FeGe with a thickness of several tens of nanometres. (**a**) Crystal structure, magnetic moment configurations of helical order, in-plane magnetic moment configuration of skyrmions when an external magnetic field B is applied downward and the corresponding over-focused LTEM images for left- and right-handed crystals. (**b**) Lateral magnetisation distribution and (**c**) under-focused image of the skyrmion crystal state obtained by LTEM observation at the boundary of left- and right-handed crystallographic domains. (**d**) Sample thickness dependence of temperature versus magnetic field phase diagram, where H, SkX and FM denote helical, skyrmion crystal and ferromagnetic spin states, respectively. The background colour indicates skyrmion density (Reproduced from (**a**) Ref. [16] and (**b**)–(**d**) Ref. [15])

bulk single crystal to obtain an observable amount of electron transmission, and a magnetic field is applied normal to the sample plane. Real-space images of lateral magnetisation distributions in helical ($B = 0$) and skyrmion crystal ($B = 50$ mT) states are shown in Fig. 2.3a, b, respectively. Consistent with the previous SANS data and the proposed spin texture shown in Fig. 2.1f, skyrmions form a triangular lattice. Here, the spin helicity, i.e. the clockwise or counter-clockwise manner of spin rotation, is fixed by the chirality of an underlying crystal via the sign of the Dzyaloshinskii–Moriya interaction in both magnetic states.

When the spin helicity changes sign, grey-scale contrasts in over- and under-focused images are reversed, as shown in Fig. 2.4a. Figure 2.4b, c indicate lateral magnetisation distributions as well as under-focused LTEM images obtained around the boundary between two opposite chiral crystallographic domains in FeGe. The spin helicity or the grey-scale contrast in under-focused images of skyrmions is clearly reversed for the opposite chiral crystallographic domain, which experimentally confirms the coupling between spin helicity and crystallographic chirality. The LTEM technique can also capture formation and annihilation processes of skyrmions, as shown in Fig. 2.3d–f. At the phase boundary between ferromagnetic and skyrmion crystal states, a magnetic skyrmion can exist as an independent defect rather than a crystallised form, as shown in Fig. 2.3f. Figure 2.3g summarises the B–T phase diagram for the thin-plate-shaped single crystal of $Fe_{1-x}Co_xSi$ ($x = 0.5$). Here, the skyrmion crystal state is stabilised down to the lowest temperature, which is in sharp contrast with the case for the bulk crystal shown in Fig. 2.1c where skyrmions appear only in a narrow temperature range just below T_c. This finding demonstrates that the stability of a skyrmion state essentially depends on the dimension or sample thickness of the system. Such a tendency is more clearly observed in the sample thickness t dependence of B–T phase diagrams for FeGe, as shown in Fig. 2.4d. The skyrmion crystal state is stable in a relatively wide B–T region for $t \sim 15$ nm, whereas it gradually shrinks for larger t and finally turns into a narrow magnetic A phase in a bulk limit [15]. When a magnetic field is applied normal to the sample plane, the helical spin order with $\mathbf{q} \parallel \mathbf{B}$ can no longer be stabilised given that the sample thickness is smaller than the helimagnetic modulation period. Such destabilisation of the helical spin state leads to the relative stabilisation of the competing skyrmion crystal state in the thin-plate-shaped sample; the possible relevance of additional uniaxial strain has also been proposed [29].

Similar magnetic phase diagrams and their sample thickness dependences are commonly observed for other B20 alloys. Despite many differences in atomic arrangement, Cu_2OSeO_3 can also be regarded as an insulating analogue of B20 alloys by considering the similarity in the crystal symmetry and magnetic phase diagram characterised by a narrow A phase (skyrmion crystal state) just below T_c (Fig. 2.1d). This implies that chiral-lattice cubic ferro/ferrimagnets may ubiquitously host skyrmion spin texture regardless of their metallic or insulating nature. The list of materials showing a skyrmion spin texture is summarised in Table 2.1. Depending on the material, various magnetic transition temperatures (up to $T_c \sim 280$ K for FeGe) and helimagnetic modulation periods (ranging from 3 nm (MnGe) to 200 nm ($Fe_{1-x}Co_xSi$)) have been reported. Note that theories predict that a skyrmion crystal can be stabilised down to $T = 0$ in the case of uniaxial non-centrosymmetric ferromagnets [1] and the emergence of magnetic skyrmions in non-centrosymmetric anti-ferromagnets has also been discussed [30]. Because various unique forms of magnetic skyrmions with different spin textures have been proposed for these systems [31], further searches for new materials and the establishment of a material design strategy are highly anticipated.

In addition to SANS for bulk samples and LTEM for thin-plate samples, the magnetic force microscopy (MFM) technique can be employed for the real-space imaging of spin textures at the surface of bulk magnetic materials with typical spatial resolutions of $\sim$20 nm [14]. Figure 2.5a, b indicate the real-space distribution of vertical magnetisation components obtained by MFM measurement under various magnitudes of a magnetic field along the out-of-plane direction for $Fe_{1-x}Co_xSi$

Table 2.1 List of materials hosting skyrmion spin texture. Magnetic ordering temperature T_c and spin modulation period λ_m are also indicated

Category	Material	T_c (K)	λ_m (nm)	Conductivity	Ref.
Chiral-lattice ferromagnets	MnSi	30	18	Metal	[6, 10]
	$Fe_{1-x}Co_xSi$	<36	$40 \sim 230$	Semiconductor	[11, 12]
	MnGe	170	3	Metal	[18]
	FeGe	278	70	Metal	[15]
	Cu_2OSeO_3	59	62	Insulator	[19, 22]
Centrosymmetric ferromagnets	$Y_3Fe_5O_{12}$	560	>500	Insulator	[32]
	$RFeO_3$	>600	>100,000	Insulator	[32]
	$BaFe_{11.79}Sc_{0.16}Mg_{0.05}O_{19}$	>300	200	Insulator	[40]
	$La_{1.37}Sr_{1.63}Mn_2O_7$	100	160	Insulator	[41]
Interface	Fe/Ir(111)	(>300)	1	Metal	[43]
	FePd/Ir(111)	(>300)	7	Metal	[44]

($x = 0.5$). The triangular lattice of skyrmions is observed at $H = 20$ mT, and they gradually turn into a helical spin texture when B is reduced. In this process, two skyrmions appear to merge into a single elongated skyrmion. The corresponding Monte-Carlo simulation suggests that the original spin texture in the skyrmion crystal state is rod-like (i.e. almost uniform along the applied B direction), as shown in Fig. 2.5c, and the transition into the helical spin state is accompanied by the merging of two skyrmion lines (Fig. 2.5d). Because this causes a change of the topological winding number, hedgehog-like point defects with finite winding number (± 1) will always appear at a merging point. Thus, the observed merger of skyrmions in MFM measurements implies that such defects pass through the sample surface. This topological point defect can be considered to be an emergent magnetic monopole or anti-monopole and will have profound effects on the dynamics of associated conduction electrons.

2.2 Skyrmions in Centrosymmetric Magnets

Next, we discuss skyrmions in centrosymmetric ferromagnets with easy-axis anisotropy [33–37]. Here, the Dzyaloshinskii–Moriya interaction is no longer relevant, and an interplay between a magnetic dipole–dipole interaction and uniaxial magnetic anisotropy causes the formation of a skyrmion spin texture. In conventional ferromagnets, the exchange interaction favours a uniform spin alignment, whereas the dipole–dipole interaction favours closed loops of magnetisation alignment. Combined with magnetic anisotropy, such competition leads to the formation of a rich variety of magnetic domain structures. When ferromagnetic domains with $\pm M$ magnetisations are separated by a domain wall parallel to the M axis (180° domain wall), two main types of spin textures can

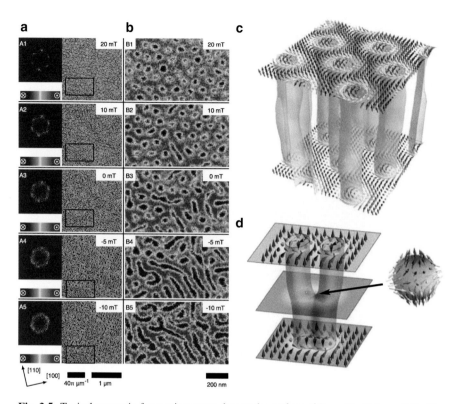

Fig. 2.5 Typical magnetic force microscopy data at the surface of $Fe_{1-x}Co_xSi$ ($x = 0.5$). *Blue* (*red*) colours correspond to magnetisation pointing parallel (anti-parallel) to the line of sight into (out of) the surface. (**a**) Data recorded as a function of a magnetic field after field cooling at $+20$ mT down to $T = 10$ K. Panel (*A1*) displays data immediately after field cooling. After the initial cool down, the field is reduced at a fixed temperature of 10 K (*A2–A5*). During this process, skyrmions, visible as *blue spots*, merge and form elongated line-like structures. The *left inset* shows the Fourier transform of a real-space signal. Magnified images at corresponding field strengths are also shown in (**b**). (**c**) Typical spin configuration of a skyrmion lattice obtained by Monte Carlo simulation. (**d**) Sketch of a magnetic configuration describing the merging of two skyrmions. Magnetisation vanishes at a singular merging point. This defect can be interpreted as an emergent magnetic anti-monopole, which acts similar to the slider of a zipper connecting two skyrmion lines (Reproduced from Ref. [14])

emerge in a domain wall region. Magnetic moments generally show a screw-like continuous rotation within a plane parallel to the domain wall, and this type of domain wall is called a 'Bloch wall' (Fig. 2.6a). In contrast, magnetic anisotropy often stabilises the cycloidal continuous rotation of magnetic moments within a plane normal to the domain wall, and this type is called a 'Néel wall' (Fig. 2.6b). Because clockwise and anti-clockwise manners of spin rotations are degenerated in centrosymmetric magnets, two types of spin helicities can randomly occur for each type of magnetic domain wall.

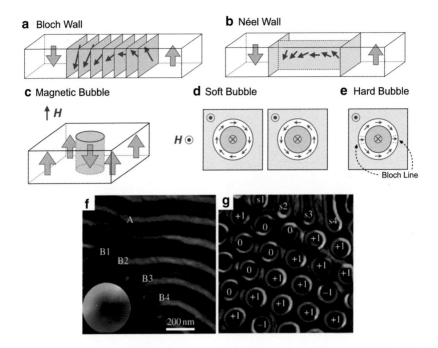

Fig. 2.6 Schematic of spin configurations at a ferromagnetic domain wall: (**a**) Bloch wall and (**b**) Néel wall, where spins rotate within a plane parallel and normal to the domain wall plane, respectively. (**c**) Magnetic bubble domain and its variation with different spin configurations at a domain wall region: (**d**) Soft bubble and (**e**) hard bubble. (**f**) and (**g**) Lateral magnetisation distribution obtained by LTEM observation for Sc-doped barium hexaferrite: (**f**) Stripe-like ferromagnetic domains and (**g**) bubble domains. The background colour indicates the direction of lateral magnetisation, and the numbers ±1 and 0 indicate the *soft bubbles* with clockwise or counter-clockwise spin rotation and the hard bubble, respectively (Reproduced from Ref. [39])

When a thin epitaxial layer of a ferromagnet has sufficiently large easy-axis magnetic anisotropy perpendicular to a film plane, the application of B normal to the film leads to the formation of a cylindrical ferromagnetic domain, as shown in Fig. 2.6c, which is called a 'magnetic bubble'. Here, the magnetisation direction at the inside (outside) of the bubble is anti-parallel (parallel) to the **B** direction, and thus, the boundary of a magnetic bubble consists of 180° domain walls. When a domain wall consists of a Bloch wall and its spin helicity is sustained throughout the cylindrical domain wall region, as shown in Fig. 2.6d, this magnetic bubble is called a 'soft bubble' and can be considered to be a type of skyrmion spin texture with a topological skyrmion number of −1. Here, unlike the case for skyrmions in chiral-lattice magnets in which the spin helicity is fixed by the crystallographic chirality, skyrmions or soft bubbles in centrosymmetric ferromagnets can have either type of spin helicity at random [38]. Under the spatial gradient of an external magnetic field, soft bubbles can be moved along the field gradient direction. Combined with the topological stability of these bubbles, the existence/absence of a magnetic bubble at a specific position can be used as information in a 0/1 bit. Such a concept has indeed evolved into a magnetic storage device called 'bubble memory', which was commercially available in the 1970s–1980s [34]. The representative material examples employed here are epitaxial films of orthoferrite $RFeO_3$ or garnet RFe_5O_{12} [32, 33].

Because the centrosymmetric ferromagnets have a spin-helicity degree of freedom that is absent in chiral-lattice ferromagnets, magnetic bubbles in these materials can be endowed with additional inner details. For example, a cylindrical domain wall often contains two types of Bloch walls with opposite spin helicities, and their boundary forms a one-dimensional line called the 'Bloch line' characterised by Néel-wall-like spin modulation. The magnetic bubble containing such Bloch lines is called a 'hard bubble' (Fig. 2.6e) and is a topologically different object from a soft bubble or a magnetic skyrmion because its skyrmion number is zero. The response of hard bubbles against an external field has been known to be different from that of soft bubbles, which often prevents the proper operation of bubble memory [38]. Figure 2.6f and g indicate typical real-space lateral magnetisation distributions for stripe-like and bubble ferromagnetic domains revealed by LTEM observation [39]. Soft bubbles with either type of spin helicity (denoted as '±1') as well as hard bubbles (denoted as '0') appear, thereby forming a triangular lattice in combination.

Apart from such hard bubbles characterised by Bloch lines, the recent advancement of the LTEM technique has also revealed several unique spin textures associated with soft bubbles, i.e. skyrmions. For example, in the case of skyrmions in M-type hexaferrite $BaFe_{12-x-0.05}Sc_xMg_{0.05}O_{19}$ ($x = 0.16$), the spin texture at a cylindrical domain wall region shows Bloch-wall-like screw magnetisation rotation but with multiple helicity reversals (Fig. 2.7) [40]. This means that the in-plane component of magnetisation changes its spin rotation manner depending on the distance from bubble centre (Fig. 2.7g), thereby giving rise to multiple rings within a single skyrmion in its LTEM image (Fig. 2.7f). Here, even with such random helicity reversals, the overall skyrmion number remains -1. Another interesting example is bilayered perovskite $La_{2-2x}Sr_{1+2x}Mn_2O_7$ [41]; in this material, two skyrmions with opposite spin helicities spontaneously make a pair and form a molecule-like structure called a biskyrmion (Fig. 2.8). When neighbouring skyrmions have the same spin helicity, the overlapping area magnetises in an anti-parallel manner over a short distance, thereby causing a large increase in exchange energy. In contrast, if their spin helicities are opposed, only a moderate modification of spin orientation is necessary in the overlapping region. This difference makes the observed biskyrmion spin texture relatively stable. Because each single-skyrmion particle has a skyrmion number of -1, the skyrmion number of a biskyrmion is -2. Biskyrmions further form an anisotropic triangular lattice spontaneously and can be driven by external electric current two orders of magnitude smaller ($<10^8$ A/m^2) than that for a conventional ferromagnetic domain wall.

As demonstrated above, the additional spin-helicity degree of freedom in centrosymmetric ferromagnets can provide a relatively richer variety of spin textures associated with magnetic skyrmions. The further investigation of their zoology as well as their interaction with an external field is the issue for the future.

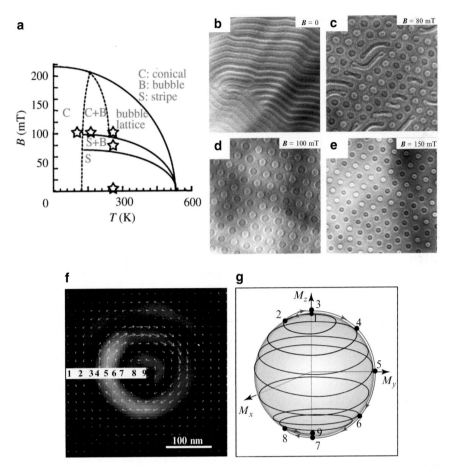

Fig. 2.7 LTEM data for $BaFe_{12-x-0.05}Sc_xMg_{0.05}O_{19}$ ($x = 0.16$). (**a**) Temperature versus magnetic field phase diagram of magnetic domain structure. (**b**)–(**e**) Magnetic field dependence of domain structure at room temperature, where B is applied normal to the observed (001) plane. (**f**) Lateral magnetisation distribution for the observed skyrmion (*soft bubble*), and (**g**) mapping of spin orientation in the skyrmion to the M_x-M_y-M_z plane (Reproduced from (**a**)–(**e**) Ref. [39] and (**f**)–(**g**) Ref. [40])

2.3   Skyrmions at Interface

As introduced in Sect. 2.1, the Dzyaloshinskii–Moriya interaction emerging under the non-centrosymmetric environment can stabilise helimagnetism and skyrmion spin texture. This scenario is valid not only for ferromagnetic materials endowed with a non-centrosymmetric crystal lattice but also for interfaces or surfaces of ferromagnets where the spatial inversion symmetry is always broken. Here, we mainly discuss two material systems: (i) Fe monolayer on an Ir(111) surface [43]

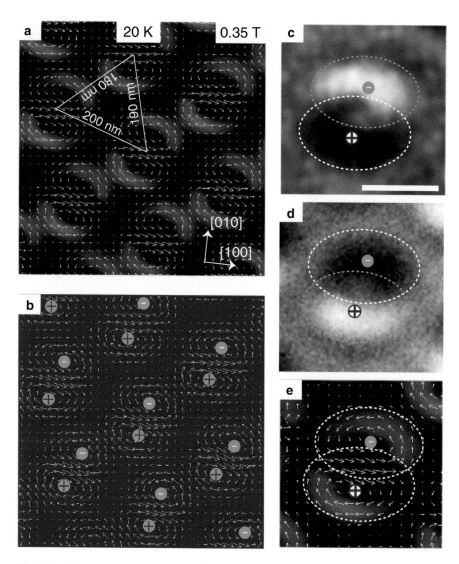

Fig. 2.8 LTEM data for $La_{2-2x}Sr_{1+2x}Mn_2O_7$. (**a**) Lateral magnetisation distribution for a biskyrmion lattice derived from the TIE analysis of LTEM data. (**b**) The in-plane magnetic component (*yellow arrows*) distribution in the biskyrmion lattice. (**c**) Over-focused and (**d**) under-focused LTEM images as well as (**e**) the corresponding lateral magnetisation distribution for a single biskyrmion. In (**b**)–(**e**), 'plus' (+) and 'minus' (−) indicate magnetic helicity, i.e. clockwise and anti-clockwise rotating directions of in-plane magnetisations around the core, respectively. Note that they are not directions of magnetisation along the z axis. The scale bar in (**c**) corresponds to 100 nm (Reproduced from Ref. [41])

and (ii) a PdFe bilayer on an Ir(111) surface [44]. The Fe (PbFe) layer provides magnetic moments with a ferromagnetic exchange interaction whose thickness is just one (two) atomic units. The large nuclear number of the underlying Ir leads to a strong spin–orbit coupling and thus can serve as a source of significant Dzyaloshinskii–Moriya interaction. Because magnitudes of relevant magnetic interactions are different between cases (i) and (ii), each system shows distinctive properties in the resultant spin texture and its development under an external field.

To investigate the detail of the spin texture at the surface of a magnetic mono/bilayer with atomic-scale spatial resolution, the spin-polarised scanning tunnelling microscopy (SP-STM) technique has been employed. The ferromagnetic tip used for this measurement is magnetised along some specific direction in advance, and spin-polarised tunnel current scales as the cosine of the angle between the tip and local sample magnetisation. As a result, this method enables the detection of a local magnetisation component parallel to the tip magnetisation direction. The pioneering work by Wiesendanger et al. for a Mn monolayer on a W(110) surface revealed the emergence of a helical spin texture of fixed spin helicity with the period of 12 nm, confirming the importance of the Dzyaloshinskii–Moriya interaction at the surface or interface [42]. They further discovered the formation of a square lattice of skyrmions (with skyrmion number $+1$) for the Fe/Ir(111) system (Fig. 2.9a) by taking SP-STM images with various directions of tip magnetisation (Fig. 2.9c, e) and comparing them with simulated patterns (Fig. 2.9d) [43]. This unique skyrmion spin texture appears even without the application of an external magnetic field and is characterised by two orthogonal magnetic modulation vectors within a film plane (see insets of Fig. 2.9c). The typical size of an individual magnetic skyrmion is as small as $\sim$1 nm, with a magnetic modulation period incommensurate with the underlying crystalline lattice. The following theoretical analysis suggests that the short-range four-spin interaction

$$\mathscr{H}_{four} = \sum_{ijkl} K_{ijkl}[(\mathbf{M}_i \cdot \mathbf{M}_j)(\mathbf{M}_k \cdot \mathbf{M}_l) + (\mathbf{M}_i \cdot \mathbf{M}_l)(\mathbf{M}_j \cdot \mathbf{M}_k) - (\mathbf{M}_i \cdot \mathbf{M}_k)(\mathbf{M}_j \cdot \mathbf{M}_l)],$$

$$(2.2)$$

as well as the Dzyaloshinskii–Moriya interaction play crucial roles in the stabilisation of the observed skyrmion lattice.

In contrast, the magnetic behaviour for the PdFe/Ir(111) system is rather close to that for the film of chiral-lattice ferromagnets, as introduced in Sect. 2.1 [44]. Without an external magnetic field, this system shows a helical spin order characterised by unidirectional spin modulation with a period of $6 \sim 7$ nm (Fig. 2.10a, d, e). The application of $H \sim 1.4$ T normal to the film leads to the formation of a hexagonal lattice of skyrmions (Fig. 2.10f) whose lateral magnetisation components are not vortical but rather radiative (Fig. 2.10b). A further increase of B ($>$2 T) stabilises the uniform ferromagnetic state, while isolated skyrmion particles often survive through pinning at atomic defects (Fig. 2.10f). Notably, when B is tuned around the phase boundary between ferromagnetic and skyrmion spin states, the local injection of electrons through a ferromagnetic tip enables the reversible

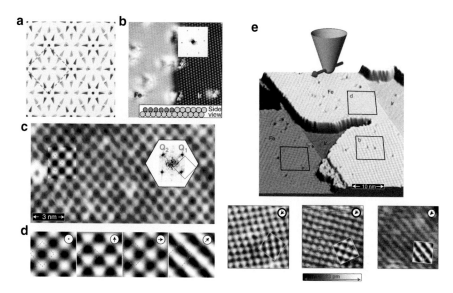

Fig. 2.9 Skyrmion lattice observed by a spin-polarised scanning tunnelling microscopy (SP-STM) measurement for the Fe monolayer on Ir(111). (**a**) Sketch of the nanoskyrmion lattice: *cones* represent atoms of a hexagonal Fe layer and point along their magnetisation directions; *red* and *green* represent up and down magnetisation components, respectively. (**b**) Atomic-resolution STM images of a pseudomorphic hexagonal Fe layer at an Ir step edge. *Upper inset*: The Fourier-transformed image. *Lower inset*: A side view of the system. (**c**) and (**e**) SP-STM images obtained with a tip magnetised along the (**c**) out-of-plane and (**e**) in-plane directions, respectively. Bright (*dark*) spots indicate areas with magnetisation parallel (anti-parallel) to tip magnetisation. In the latter case, images corresponding to three 120° rotational domains are obtained for different square regions shown in the *upper panel*. The tip magnetisation direction is indicated by *arrows*. In (**c**), the Fourier transformation of the experimental SP-STM image is also indicated. (**d**) Simulations of SP-STM images of a skyrmion lattice with the tip magnetised in different directions as indicated. Here, the image size and unit-cell position are identical to those in (**a**). Simulated patterns are also overlaid on experimentally obtained images in (**c**) and (**e**), which agree well with each other (Reproduced from Ref. [43])

writing and deleting of single magnetic skyrmions (Fig. 2.10h, i). The probability of skyrmion formation after current injection strongly depends on the sign of applied electric current, suggesting that the spin-transfer torque provided by spin-polarised tunnelling current affects the directionality of switching. The above procedures enable the local 'writing' and 'reading' of individual skyrmion particles of an extremely small size, i.e. less than 10 nm, which demonstrates the potential of magnetic skyrmions as an information carrier for high-density storage/logic devices.

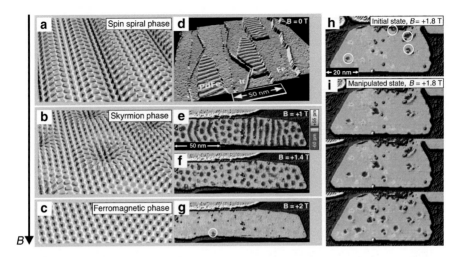

Fig. 2.10 Magnetic field dependence of a PdFe bilayer on the Ir(111) surface at 8 K. (**a**)–(**c**) Perspective sketches of magnetic phases. (**d**) Overview image obtained by SP-STM, perspective view of constant-current image colorized with its derivative. (**e**)–(**g**) Distribution of the out-of-plane component of magnetisation for a PdFe bilayer at different magnetic fields along the out-of-plane direction. (**e**) Coexistence of spin spiral and skyrmion phases. (**f**) Pure skyrmion phase. (**g**) Ferromagnetic phase. A remaining skyrmion is marked by the *white circle*. In (**h**) and (**i**), the manipulation of magnetic states by a local current injection through a magnetic tip is demonstrated at 4.2 K. (**h**) SP-STM image of the initial state at B = +1.8 T after sweeping the magnetic field down from +3 T. Four skyrmions are marked by *circles*. (**i**) Successive population of an island with skyrmions by injecting higher-energy electrons through local voltage sweeps (Reproduced from Ref. [44])

References

1. A.N. Bogdanov, D.A. Yablonskii, Sov. Phys. JETP **68**, 101 (1989)
2. A. Bogdanov, A. Hubert, J. Magn. Magn. Mat. **138**, 255 (1994)
3. U.K. Rößler, A.N. Bogdanov, C. Pfleiderer, Nature **442**, 797 (2006)
4. I. Dzyaloshinskii, J. Phys. Chem. Solids **4**, 241 (1958)
5. T. Moriya, Phys. Rev. **120**, 91 (1960)
6. S. Mühlbauer, B. Binz, F. Jonietz, C. Pfleiderer, A. Rosch, A. Neubauer, R. Georgii, P. Böni, Science **323**, 915 (2009)
7. C. Pappas, E. Lelièvre-Berna, P. Falus, P.M. Bentley, E. Moskvin, S. Grigoriev, P. Fouquet, B. Farago, Phys. Rev. Lett. **102**, 197202 (2009)
8. C. Pfleiderer, T. Adams, A. Bauer, W. Biberacher, B. Binz, F. Birkelbach, P. Böni, C. Franz, R. Georgii, M. Janoschek, F. Jonietz, T. Keller, R. Ritz, S. Mühlbauer, W. Munzer, A. Neubauer, B. Pedersen, A. Rosch, J. Phys. Condens. Matter **22**, 164207 (2010)
9. T. Adams, S. Mühlbauer, C. Pfleiderer, F. Jonietz, A. Bauer, A. Neubauer, R. Georgii, P. Böni, U. Keiderling, K. Everschor, M. Garst, A. Rosch, Phys. Rev. Lett. **107**, 217206 (2011)
10. A. Tonomura X.Z. Yu, K. Yanagisawa, T. Matsuda, Y. Onose, N. Kanazawa, H.S. Park, Y. Tokura, Nano Lett. **12**, 1673 (2012)
11. W. Munzer, A. Neubauer, T. Adams, S. Mühlbauer, C. Franz, F. Jonietz, R. Georgii, P. Böni, B. Pedersen, M. Schmidt, A. Rosch, C. Pfleiderer, Phys. Rev. B **81**, 041203(R) (2010)

12. X.Z. Yu, Y. Onose, N. Kanazawa, J.H. Park, J.H. Han, Y. Matsui, N. Nagaosa, Y. Tokura, Nature **465**, 901 (2010)
13. D. Morikawa, K. Shibata, N. Kanazawa, X.Z. Yu, Y. Tokura, Phys. Rev. B **88**, 024408 (2013)
14. P. Milde, D. Köhler, J. Seidel, L.M. Eng, A. Bauer, A. Chacon, J. Kindervater, S. Mühlbauer, C. Pfleiderer, S. Buhrandt, C. Schüte, A. Rosch, Science **340**, 1076 (2013)
15. X.Z. Yu, N. Kanazawa, Y. Onose, K. Kimoto, W.Z. Zhang, S. Ishiwata, Y. Matsui, Y. Tokura, Nat. Mater. **10**, 106 (2011)
16. K. Shibata, X.Z. Yu, T. Hara, D. Morikawa, N. Kanazawa, K. Kimoto, S. Ishiwata, Y. Matsui, Y. Tokura, Nat. Nanotech. **8**, 723 (2013)
17. S.V. Grigoriev, N.M. Potapova, S.-A. Siegfried, V.A. Dyadkin, E.V. Moskvin, V. Dmitriev, D. Menzel, C.D. Dewhurst, D. Chernyshov, R.A. Sadykov, L.N. Fomicheva, A.V. Tsvyashchenko, Phys. Rev. Lett. **110**, 207201 (2013)
18. N. Kanazawa, Y. Onose, T. Arima, D. Okuyama, K. Ohoyama, S. Wakimoto, K. Kakurai, S. Ishiwata, Y. Tokura, Phys. Rev. Lett. **106**, 156603 (2011)
19. S. Seki, X.Z. Yu, S. Ishiwata, Y. Tokura, Science **336**, 198 (2012)
20. S. Seki, J.-H. Kim, D.S. Inosov, R. Georgii, B. Keimer, S. Ishiwata, Y. Tokura, Phys. Rev. B **85**, 220406 (2012)
21. S. Seki, S. Ishiwata, Y. Tokura, Phys. Rev. B **86**, 060403 (2012)
22. T. Adams, A. Chacon, M. Wagner, A. Bauer, G. Brandl, B. Pedersen, H. Berger, P. Lemmens, C. Pfleiderer, Phys. Rev. Lett. **108**, 237204 (2012)
23. P. Bak, M.H. Jensen, J. Phys. C **13**, L881 (1980)
24. K. Kadowaki, K. Okuda, M. Date, J. Phys. Soc. Jpn. **51**, 2433 (1982)
25. M. Date, K. Okuda, K. Kadowaki, J. Phys. Soc. Jpn. **42**, 1555 (1977)
26. S. Kusaka, K. Yamamoto, T. Komatsubara, Y. Ishikawa, Solid State Commun. **20**, 925 (1976)
27. A. Neubauer, C. Pfleiderer, B. Binz, A. Rosch, R. Rtiz, P.G. Niklowitz, P. B oni, Phys. Rev. Lett. **102**, 186602 (2009)
28. M. Uchida, Y. Onose, Y. Matsui, Y. Tokura, Science **311**, 359 (2006)
29. A.B. Butenko, A.A. Leonov, U.K. Rößler, A.N. Bogdanov, Phys. Rev. B **82**, 052403 (2010)
30. A.N. Bogdanov, U.K. Rößler, M. Wolf, K.-H. Müller, Phys. Rev. B **66**, 214410 (2002)
31. A.B. Butenko, A.A. Leonov, A.N. Bogdanov, U.K. Rößler, J. Phys. Conf. Ser. **200**, 042012 (2009)
32. E.A. Giess, Science **208**, 938 (1980)
33. Y.S. Lin, J. Grundy, E.A. Giess, Phys. Lett. **23**, 485 (1973)
34. A.P. Malozemoff, J.C. Slonczewski, *Magnetic Domain Walls in Bubble Materials* (Academic, New York, 1979), pp. 306–314
35. T. Garel, S. Doniach, Phys. Rev. B **26**, 325 (1982)
36. T. Suzuki, J. Magn. Magn. Mater. **31–34**, 1009 (1983)
37. A. Hubert, R. Schäfer, *Magnetic Domains* (Springer, Berlin/New York, 1998)
38. A Correspondent, Nature **240**, 184 (1972)
39. N. Nagaosa, X.Z. Yu, Y. Tokura, Phil. Trans. R. Soc. A **370**, 5806 (2012)
40. X.Z. Yu, M. Mostovoy, Y. Tokunaga, W. Zhang, K. Kimoto, Y. Matsui, Y. Kaneko, N. Nagaosa, Y. Tokura, Proc. Natl. Acad. Sci. U.S.A. **109**, 8856 (2012)
41. X.Z. Yu, Y. Tokunaga, Y. Kaneko, W.Z. Zhang, K. Kimoto, Y. Matsui, Y. Taguchi, Y. Tokura, Nat. Commun. **5**, 4198 (2014)
42. S. Heinze, M. Bode, A. Kubetzka, O. Pietzsch, X. Nie, S. Blügel, R. Wiesendanger, Science **288**, 1805 (2000)
43. S. Heinze, K. von Bergmann, M. Menzel, J. Brede, A. Kubetzka, R. Wiesendanger, G. Bihlmayer, S. Blügel, Nat. Phys. **7**, 713 (2011)
44. N. Romming, C. Hanneken, M. Menzel, J.E. Bickel, B. Wolter, K. von Bergmann, A. Kubetzka, R. Wiesendanger, Science **341**, 636 (2013)

Chapter 3
Skyrmions and Electric Currents in Metallic Materials

Abstract In metallic materials, non-collinear or non-coplanar spin textures such as skyrmions, helices or domain walls give rise to intriguing phenomena via coupling to conduction electrons. In this chapter, we introduce the emergent electromagnetic fields generated by a skyrmion spin texture acting on conduction electrons. These cause the topological Hall effect and the electric-current-driven motion of skyrmions with a significantly small threshold current density j_c of 10^5–10^6 A/m^2, which is five or six orders of magnitude smaller than that of a ferromagnetic domain wall and a helical magnetic structure.

3.1 Emergent Electromagnetic Fields

In metallic magnets, spatially varying structures of localised magnetisation generate emergent magnetic and electric fields acting on conduction electrons via coupling between localised spins and conduction-electron spins [1], thereby giving rise to interesting electron transport phenomena [2, 3]. The simplest model to describe this situation is given by

$$i\hbar \frac{\partial}{\partial t} \Psi = \left[\frac{p^2}{2m_e} - J_{ex} \boldsymbol{\sigma} \cdot \mathbf{m}(\mathbf{r}, t) \right] \Psi, \qquad (3.1)$$

where the Pauli matrix $\boldsymbol{\sigma}$ denotes the conduction-electron spin and the unit vector $\mathbf{m}(\mathbf{r}, t)$ represents the direction of local magnetisation. The magnetisation vector $\mathbf{m}(\mathbf{r}, t)$ is given in polar coordinates as

$$\mathbf{m}(\mathbf{r}, t) = (\sin \theta(\mathbf{r}, t) \cos \phi(\mathbf{r}, t), \quad \sin \theta(\mathbf{r}, t) \sin \phi(\mathbf{r}, t), \quad \cos \theta(\mathbf{r}, t)). \qquad (3.2)$$

Here, a quantisation axis is selected to be parallel to the z axis. When a ferromagnetic (anti-ferromagnetic) coupling J_{sd} or the Hund's rule coupling J_H is sufficiently strong, the conduction-electron spin always points parallel (anti-parallel) to localised magnetisation $\mathbf{m}(\mathbf{r}, t)$ as it flows and follows the spatial and temporal variations of $\mathbf{m}$. On the basis of this assumption, we rotate the local

© Springer International Publishing Switzerland 2016

S. Seki, M. Mochizuki, *Skyrmions in Magnetic Materials*, SpringerBriefs in Physics, DOI 10.1007/978-3-319-24651-2_3

quantisation axis from the fixed axis $\mathbf{e}_z$ to the axis parallel to $\mathbf{m}$ at a given $(\mathbf{r}, t)$ by introducing

$$\Psi = U(\mathbf{r}, t)\varphi \tag{3.3}$$

with

$$U = \exp(-i\frac{\theta}{2}\boldsymbol{\sigma} \cdot \mathbf{n}), \tag{3.4}$$

where θ is the angle of rotation, and the axis of the rotation $\mathbf{n}$ is

$$\mathbf{n} = \frac{\mathbf{e}_z \times \mathbf{m}}{|\mathbf{e}_z \times \mathbf{m}|}. \tag{3.5}$$

Substituting Eq. (3.3) into Eq. (3.1), we obtain

$$i\hbar\frac{\partial}{\partial t}\varphi = \left[\frac{(\mathbf{p} + e\mathbf{A}^s)^2}{2m_e} - J_{\text{ex}}\sigma_z - eV^s\right]\varphi, \tag{3.6}$$

where $e(> 0)$ is the elementary charge. The vector potential $\mathbf{A}^s$ is given by

$$\mathbf{A}^s = -\frac{i\hbar}{e}U^\dagger\nabla U, \tag{3.7}$$

and the scalar potential V^s is given by

$$V^s = \frac{i\hbar}{e}U^\dagger\partial_t U. \tag{3.8}$$

As long as the spin texture $\mathbf{m}(\mathbf{r}, t)$ is slowly varying in space and time, the vector potential and scalar potential can be treated as perturbations of the unperturbed Hamiltonian:

$$\mathcal{H}_0 = \frac{p^2}{2m_e} - J_{\text{ex}}\sigma_z. \tag{3.9}$$

This Hamiltonian describes the spin-up band with respect to the local magnetisation direction.

Effective magnetic and electric fields can be introduced as

$$B_i^{\text{em}} = \epsilon_{ijk}\left(\partial_j A_k^s - \partial_k A_j^s\right) = \frac{\hbar}{2e}\epsilon_{ijk}\ \mathbf{m} \cdot \left(\partial_j\mathbf{m} \times \partial_k\mathbf{m}\right), \tag{3.10}$$

and

$$E_i^{\text{em}} = -\partial_i V^s - \partial_t A_i^s = \frac{\hbar}{e}\mathbf{m} \cdot \left(\partial_i\mathbf{m} \times \partial_t\mathbf{m}\right), \tag{3.11}$$

respectively, with $(\partial_i, \partial_j, \partial_k)=(\partial/\partial x, \partial/\partial y, \partial/\partial z)$ and $\partial_t=\partial/\partial t$. Here, $\pm$ denotes the spin-up or spin-down bands and ϵ_{ijk} is the totally anti-symmetric tensor. These expressions indicate that these emergent magnetic and electric fields are associated with the Berry phase as the effective Aharonov–Bohm phase given by solid angles covered by $\mathbf{m}$ for infinitesimal loops in space and space-time dimensions, respectively.

Because the magnetisation $\mathbf{m}$ of a single skyrmion winds around the unit sphere once, the total flux is given by

$$\int \mathbf{B}^{em} dx dy = -4\pi \frac{\hbar}{2e} = -\frac{h}{2q_e}, \tag{3.12}$$

which corresponds to one flux quantum with $q_e(= e/2)$ being the effective charge.

In analogy with Faraday's law of induction, a skyrmion drifting with velocity $\mathbf{v}_d$ induces an electric field given by

$$\mathbf{E}^{em} = -\mathbf{v}_d \times \mathbf{B}^{em}, \tag{3.13}$$

where E^{em}/v_d inherits its quantisation from B^{em}.

The total force acting on a conduction electron with velocity $\mathbf{v}_s$ is therefore given by

$$\mathbf{F} = -e(\mathbf{E}^{ex} + \mathbf{E}^{em}) - e(\mathbf{v}_s - \mathbf{v}_d) \times (\mathbf{B}^{ex} + \mathbf{B}^{em}), \tag{3.14}$$

where $\mathbf{B}^{ex}$ and $\mathbf{B}^{em}$ ($\mathbf{E}^{ex}$ and $\mathbf{E}^{em}$) are external and emergent magnetic (electric) fields, respectively, and $\mathbf{v}_s$ and $\mathbf{v}_d$ are velocities of an electron and a moving skyrmion. The Lorentz force $-e\mathbf{v}_s \times \mathbf{B}^{em}$ due to the emergent magnetic field $\mathbf{B}^{em}$ gives rise to a Hall motion of conduction electrons, which is called the topological Hall effect. Notably, because the relation Eq. (3.13) holds, contributions $-e\mathbf{E}^{em}$ and $+e\mathbf{v}_d \times \mathbf{B}^{ex}$ in Eq. (3.14) are perfectly cancelled.

3.2 Electric-Current-Driven Motions of Skyrmions

The magnetisation dynamics driven by spin-polarised electric current is described by the Landau–Lifshitz–Gilbert–Slonczewski equation [4] as follows:

$$\frac{d\mathbf{M}}{dt} = -\gamma \mathbf{M} \times \mathbf{B}_r^{eff} + \frac{\alpha_G}{M} \mathbf{M} \times \frac{d\mathbf{M}}{dt}$$
$$+ \frac{pa^3}{2e} \frac{\gamma\hbar}{M} (\mathbf{j} \cdot \nabla)\mathbf{M} - \frac{pa^3\beta}{2e} \frac{\gamma\hbar}{M^2} [\mathbf{M} \times (\mathbf{j} \cdot \nabla)\mathbf{M}], \tag{3.15}$$

where $\mathbf{M}(\mathbf{r})=-\gamma\mathbf{S}(\mathbf{r})$ represents local magnetisation at position $\mathbf{r}$ and $\mathbf{S}(\mathbf{r})$ is the local spin. Here, $\gamma = g\mu_B/\hbar(> 0)$ and $e(> 0)$ are the gyromagnetic ratio and elementary charge, respectively. This equation describes the magnetisation

dynamics in the presence of spin-polarised electric current $\mathbf{j}$ [5]. The first term
(the gyrotropic term) depicts the gyrotropic motion of $\mathbf{M}$, where the effective
magnetic field $\mathbf{B}^{\text{eff}}$ is calculated from a derivative of the Hamiltonian $\mathscr{H}$ with
respect to $\mathbf{M}$ as

$$\mathbf{B}^{\text{eff}} = -\frac{\partial \mathscr{H}}{\partial \mathbf{M}}. \tag{3.16}$$

The second term (the Gilbert damping term) describes the phenomenologically
introduced Gilbert damping, whose strength is represented by the coefficient α_{G}.
The third and fourth terms represent the coupling between magnetisations $\mathbf{M}$ and
spin-polarised electric current $\mathbf{j}$. Here, p and a are the spin polarisation of electric
current and the lattice constant, respectively. Microscopically, conduction-electron
spins interact with $\mathbf{M}$ via a local exchange interaction J_{sd} or the Hund's rule coupling
J_{H}. Because spin-polarised electric current has a flux of angular momentum, it works
as a torque acting on $\mathbf{M}$ in non-collinear spin structures. The third term, the so-called
spin-transfer-torque term, is derived under the assumption that conduction-electron
spins are always parallel to $\mathbf{M}$ in the limit of strong J_{H} and J_{sd}. On the other hand, the
fourth term, the so-called non-adiabatic or β term, represents the coupling between
$\mathbf{M}$ and spin-polarised electric current via the non-adiabatic effect whose strength is
represented by the coefficient β.

To solve Eq. (3.15) numerically, the spatial gradient ∇ in the third and fourth
terms should be defined on a discretised lattice space. In addition, because the
time derivative $\frac{d\mathbf{M}}{dt}$ appears in both the left- and right-hand sides of the equation,
a linearlisation of the equation is needed to apply numerical techniques such as
the Runge–Kutta and Heun methods. After discretisation and linearlisation, the
equation is rewritten in a dimensionless form as

$$\frac{d\mathbf{m}_i}{d\tau} = \frac{1}{1+\alpha_{\text{G}}^2}\left\{ -\mathbf{m}_i \times \left(-\frac{\partial \tilde{\mathscr{H}}}{\partial \mathbf{m}_i}\right) + \left(\tilde{\mathscr{A}}_i + \tilde{\mathscr{B}}_i\right) \right.$$
$$\left. +\frac{\alpha_{\text{G}}}{m}\mathbf{m}_i \times \left[-\mathbf{m}_i \times \left(-\frac{\partial \tilde{\mathscr{H}}}{\partial \mathbf{m}_i}\right) + \left(\tilde{\mathscr{A}}_i + \tilde{\mathscr{B}}_i\right), \right] \right\} \tag{3.17}$$

where

$$\tilde{\mathscr{A}}_i = \frac{pa^2\hbar}{2emJ}\left(j_x\frac{\mathbf{m}_{i+\hat{x}}-\mathbf{m}_{i-\hat{x}}}{2} + j_y\frac{\mathbf{m}_{i+\hat{y}}-\mathbf{m}_{i-\hat{y}}}{2}\right),$$
$$\tilde{\mathscr{B}}_i = -\frac{pa^2\hbar}{2emJ}\frac{\beta}{m}\left[\mathbf{m}_i \times \left(j_x\frac{\mathbf{m}_{i+\hat{x}}-\mathbf{m}_{i-\hat{x}}}{2} + j_y\frac{\mathbf{m}_{i+\hat{y}}-\mathbf{m}_{i-\hat{y}}}{2}\right)\right]. \tag{3.18}$$

Here, $\mathbf{m}_i = \mathbf{M}_i/\gamma\hbar = -\mathbf{S}_i/\hbar$ is dimensionless local magnetisation at the ith site
and $\tau = tJ/\hbar$ and $\tilde{\mathscr{H}} = \mathscr{H}/J$ are the dimensionless time and the Hamiltonian,
respectively. The details of the derivation are presented in the appendix of this
chapter.

To describe the magnetic system in a thin-film specimen of a chiral-lattice magnet with distributed magnetic impurities, we employ a classical Heisenberg model on the square lattice:

$$\mathcal{H} = -J \sum_i \mathbf{m}_i \cdot (\mathbf{m}_{i+\hat{x}} + \mathbf{m}_{i+\hat{y}})$$

$$-D \sum_i (\mathbf{m}_i \times \mathbf{m}_{i+\hat{x}} \cdot \hat{x} + \mathbf{m}_i \times \mathbf{m}_{i+\hat{y}} \cdot \hat{y})$$

$$-B \sum_i m_{iz} - A \sum_{i \in I} m_{iz}^2. \tag{3.19}$$

The last term represents magnetic anisotropy with its easy magnetisation axis $(A > 0)$ perpendicular to the plane at randomly distributed impurity sites. Here, I denotes a set of impurity positions. As discussed in Chap. 1, the model without the last term reproduces the successive emergence of helical, skyrmion crystal and ferromagnetic phases as a function of magnetic field B in agreement with the experiment. Note that a helical phase can be regarded as a sequence of ferromagnetic Bloch walls, enabling us to compare the current-driven motion of skyrmions on equal footing with that of magnetic domain walls by varying the strength of B without changing any other parameters.

Figure 3.1a displays simulated velocities $v_\parallel$ (the components parallel to $\mathbf{j}$) of a skyrmion crystal and helical spin structure as functions of electric current density j for several values of β, i.e. $\beta = 0, 0.5\alpha_G, \alpha_G$ and $2\alpha_G$ [6]. Both the clean case without impurities $(x = 0)$ and dirty case with impurities $(x = 0.1\%)$ are examined, with x being the impurity concentration. Remarkably, the current–velocity $(j$-$v_\parallel)$ relation for a skyrmion crystal, represented by blue and light-blue data points, is quite universal and independent of the non-adiabatic effect β, the Gilbert damping α_G and impurities. We find that all plots overlap within the accuracy of the numerical simulation.

In contrast, the j-$v_\parallel$ relation for a helical structure, represented by red and purple data points, depends sensitively on these three factors, similarly to the case of the single ferromagnetic domain wall. A helical structure cannot move when $\beta = 0$ because it is prevented by the intrinsic pinning effect. With a finite β, j-$v_\parallel$ characteristics nearly obey the relation $v_\parallel \propto (\beta/\alpha_G)j$ in the clean case with $x = 0$. In the presence of impurities, the pinning effect suppresses the velocity $v_\parallel$ and a finite threshold current density j_c appears, whose order is 10^{10}–10^{11} A/m^2 when $x = 0.1\%$ in the present model.

A reason why skyrmions are scarcely pinned by impurities is their flexibility in shape and their particle-like nature. The triangular form of a skyrmion crystal and each individual skyrmion can be flexibly deformed during motion, which enables them to move in such a manner as to avoid pinning centres. Figure 3.1b shows a trajectory of one moving skyrmion and depicts the skyrmion as a particle-like object, winding its trajectory to avoid impurities indicated by green dots.

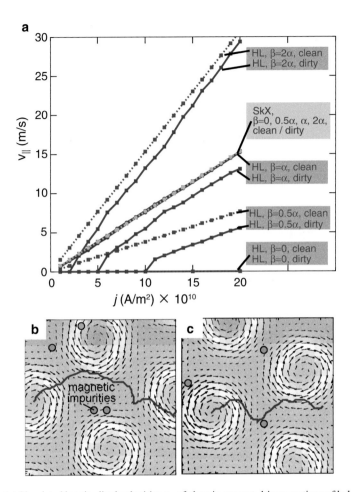

Fig. 3.1 (a) Simulated longitudinal velocities $v_\parallel$ of electric-current-driven motions of helical (*HL*) and skyrmion crystal (*SkX*) spin structures as functions of electric current density j for several values of β. Parameters $J = 1\,\mathrm{meV}$, $D/J = -0.18$, $A/J = 0.2$, $p = 0.2$ and $C\alpha_G = 0.01$ are used in the simulation, and both the clean case without impurity ($x = 0$) and dirty case with impurities ($x = 0.1\,\%$) are examined, where x is the impurity concentration. *Red* and *purple points* and *lines* represent the data of the HL structure, whereas *blue* and *light-blue points* and *lines* represent the data of the SkX structure. All lines for an SkX overlap within the accuracy of numerical simulations. (**b**) One example of a simulated trajectory of a skyrmion in a skyrmion crystal during electric-current-driven motion. The skyrmion moves to avoid impurities (*green dots*). (**c**) Another example of a simulated skyrmion trajectory. Because a skyrmion in the moving skyrmion crystal is pushed by other surrounding skyrmions, a situation where a skyrmion cannot avoid impurity sites sometimes arises. In such a case, a skyrmion rushes to the impurity site so as to let its core run over it because core magnetisation pointing downwards is also energetically favourable for magnetic anisotropy with an easy axis perpendicular to the plane (Reproduced from Ref. [6])

The topological nature of a skyrmion is also of crucial importance for the universal j-v relation with less influence from the non-adiabatic effect β, the Gilbert damping α_G and impurities. This can be understood as follows [6].

The centre-of-mass motion of a rigid spin texture is described by Thiele's equation [7], which is derived from the Landau–Lifshitz–Gilbert–Slonczewski equation by assuming that the spin texture never deforms during its drift motion. The equation is given by [8, 28] as

$$\mathbf{G} \times (\mathbf{v}_s - \mathbf{v}_d) + \mathscr{D}(\beta \mathbf{v}_s - \alpha_G \mathbf{v}_d) + \mathbf{F}_{pin} - \nabla U = \mathbf{0}, \tag{3.20}$$

where $\mathbf{v}_d$ is the drift velocity of the spin texture and $\mathbf{v}_s$ is the velocity of conduction electrons. The first term on the left-hand side describes the Magnus force, whereas the second term depicts the dissipative force. The third term $\mathbf{F}_{pin}$ denotes the phenomenologically introduced force due to the impurity pinning [8, 28]:

$$\mathbf{F}_{pin} \sim -4\pi v_{pin} f(v_d/v_{pin}) \mathbf{v}_d / |v_d|. \tag{3.21}$$

Here, f is a scaling function and $\mathbf{v}_{pin}$ is a velocity characterising the pinning strength. The last term represents a force due to the potential from the surrounding environment. The gyromagnetic coupling vector $\mathbf{G} = (0, 0, \mathscr{G})$ is given by

$$\mathscr{G} = \int_{\text{unit cell}} d^2 r \left(\frac{\partial \hat{\mathbf{n}}}{\partial x} \times \frac{\partial \hat{\mathbf{n}}}{\partial y} \right) \cdot \hat{\mathbf{n}} = 4\pi Q, \tag{3.22}$$

where $Q(= \pm 1)$ is the skyrmion number and $\hat{\mathbf{n}} = \mathbf{M}(\mathbf{r})/M$. On the other hand, components of the dissipative force tensor $\mathscr{D}$ are given by

$$\mathscr{D}_{ij} = \int_{\text{unitcell}} d^2 r \partial_i \hat{\mathbf{n}} \cdot \partial_j \hat{\mathbf{n}} = \begin{cases} \mathscr{D} & (i,j) = (x,x), (y,y), \\ 0 & \text{otherwise.} \end{cases} \tag{3.23}$$

The details of the derivation are presented in the appendix of this chapter. Importantly, the first term of Eq. (3.20) contains the topological number $\mathscr{G}$. The crucial difference between a skyrmion and a helix is the value of $\mathscr{G}$. It is $\pm 4\pi$ for a single skyrmion but zero for a helix and a domain wall. Because values of $\alpha_G (\sim 10^{-2})$ and $\beta (\sim \alpha_G)$ are much smaller than unity, the second term of Eq. (3.20) becomes negligible if $|\mathscr{G}| = 4\pi$ and the electric-current-driven motion is governed by the first term. Then, the motion of a skyrmion is well described by

$$\mathbf{G} \times (\mathbf{v}_s - \mathbf{v}_d) \sim -\mathbf{F}_{pin}. \tag{3.24}$$

From this equation, we find that the skyrmion motion is not affected by values of β and α_G. In particular, when the impurity pinning is absent or sufficiently weak ($\mathbf{F}_{pin} \sim 0$), Eq. (3.24) gives

$$\mathbf{v}_\parallel = \mathbf{v}_s, \tag{3.25}$$

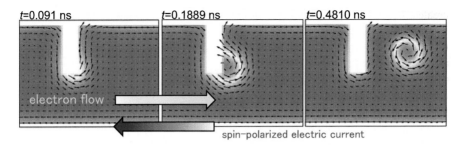

t=0.091 ns t=0.1889 ns t=0.4810 ns

electron flow

spin–polarized electric current

Fig. 3.2 Snapshots of magnetisation distribution during skyrmion creation with an electric current of $j = 3.6 \times 10^{11}$ A/m^2. Parameters used for the simulation are $J = 1$ meV, $D/J = -0.18$, $B = 0.0278J$ and $\alpha = 0.04$ (Reproduced from Ref. [10])

for the electric-current-driven motion of skyrmions in agreement with the β-insensitive universal j–v relation obtained by the numerical simulation in the clean case.

On the other hand, when $\mathscr{G} = 0$, the electric-current-driven motion is governed by the second term of Eq. (3.20). Hence, motions of helices and ferromagnetic domain walls are well described by

$$\mathscr{D}(\beta \mathbf{v}_s - \alpha_G \mathbf{v}_d) \sim -\mathbf{F}_{\text{pin}}. \tag{3.26}$$

This is the reason why the motion of a helix strongly depends on β and α_G. When $\mathbf{F}_{\text{pin}} \sim 0$, Eq. (3.26) gives

$$\mathbf{v}_{\parallel} = \frac{\beta}{\alpha_G} \mathbf{v}_s \propto \frac{\beta}{\alpha_G} \mathbf{j}. \tag{3.27}$$

This relation is again in agreement with the simulated results for the HL state in the clean case.

The numerical simulation of the Landau–Lifshitz–Gilbert–Slonczewski equation also demonstrates that skyrmions can be created with spin-polarised electric currents [10]. As shown in Fig. 3.2, by injecting electric current into a stripline-shaped sample with a small rectangular notch, one can create isolated skyrmions successively. Here, the strength of the external magnetic field B is fixed at $B = 0.0278J$, which slightly exceeds the critical field for the phase transition between a skyrmion crystal and ferromagnetic phases, where the ground state of the system is ferromagnetic. In bulk specimens, topological spin textures cannot be created by continuous twisting or gradual variation of the spatial distribution of magnetisations starting from a uniformly-polarised ferromagnetic state. One needs to flip local magnetisation to change the topological number from zero to 4π and create a skyrmion, which costs a large amount energy whose order is in J. However, taking advantage of discontinuity in the distribution of magnetisations at the notch, one can avoid such a high-energy-consuming procedure and easily create a topological spin texture. In such a process, skyrmions can be created only with electric current flowing in a certain direction determined by the sign of the magnetic field because magnetisation precession occurs in a certain direction. It is also demonstrated that one can eliminate skyrmions by colliding them against an edge of the sample through electric current injection. Here, the discontinuity of magnetisation distribution at the edge is again harnessed to change the topological skyrmion number from 4π to zero.

3.3 Topological Hall Effect

Herein, we introduce the experimental detection of emergent electromagnetic fields associated with conduction electrons interacting with the skyrmion spin texture. As discussed in Sect. 3.1, conduction electrons passing through a non-coplanar spin texture characterised by the non-zero scalar spin chirality $\mathbf{S}_i \cdot (\mathbf{S}_j \times \mathbf{S}_k)$ gain a quantum-mechanical Berry phase, which often acts as an emergent magnetic field and gives rise to additional contributions to the Hall resistivity or conductivity. This phenomenon is called the topological Hall effect. To obtain the finite topological Hall resistivity due to a non-coplanar spin texture, (i) a crystallographic lattice with a non-trivial geometry characterised by multiple inequivalent loops in the unit cell or (ii) a special topology of the spin texture characterised by a non-zero skyrmion number is required [13]. The topological Hall effect due to mechanism (i) was reported in pyrochlore $Nd_2Mo_2O_7$ [14], where the emergent magnetic field $\mathbf{B}_k^{em}$ in the momentum space provided anomalous velocity $\mathbf{v} = (e/\hbar)\mathbf{E} \times \mathbf{B}_k^{em}$ and thus an additional contribution to the Hall conductivity $\sigma_{xy}^T \propto \sum f(\epsilon) B_{kz}^{em}$. Here, $f(\epsilon)$ is the Fermi distribution function and the summation is taken over relevant bands. In contrast, the topological Hall effect due to the latter scheme (ii) is induced by the emergent magnetic field $\mathbf{B}_r^{em}$ in real space. It gives a fictitious Lorentz force $F = -e\mathbf{v}_s \times \mathbf{B}_r^{em}$, with $\mathbf{v}_s$ being electron velocity and thus an additional contribution to the Hall resistivity $\rho_{yx}^T \propto B_r^{em}$. This latter process (ii) becomes dominant when the spin modulation period λ_m is much larger than the crystallographic lattice constant (a), which corresponds to the case for B20 compounds with $a \sim 0.5\,nm$ and a skyrmion spin texture of $30\,nm < \lambda_m < 200\,nm$.

Figure 3.3b, c indicate the magnetic field dependence of the Hall resistivity ρ_{xy} measured for a bulk MnSi sample at various temperatures close to T_c [12]. For this system, ρ_{xy} can be decomposed into three contributions:

$$\rho_{xy} = \rho_{xy}^N + \rho_{xy}^A + \rho_{xy}^T \tag{3.28}$$

$$= R_0 B + S_A \rho_{xx}^2 M + P R_0 B_{rz}^{em}, \tag{3.29}$$

where ρ_{xy}^N, ρ_{xy}^A and ρ_{xy}^T correspond to the normal Hall term proportional to the external magnetic field B, the anomalous Hall term proportional to magnetisation M and the topological Hall term proportional to the emergent magnetic field B_{rz}^{em}.

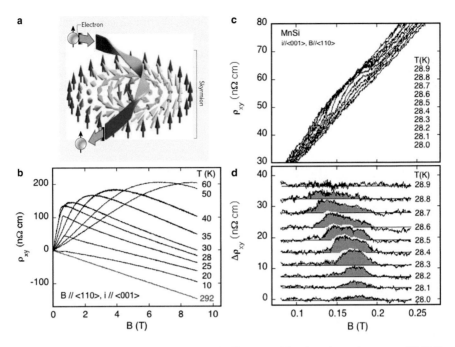

Fig. 3.3 (**a**) Schematic of the topological Hall effect caused by skyrmion spin texture. (**b**) Hall resistivity for single-crystal MnSi, where a magnetic field is applied parallel to [110] and current is applied along [001]. (**c**) Hall resistivity ρ_{xy} near T_c in the temperature and field range of the A phase. (**d**) Additional Hall contribution $\Delta\rho_{xy}$ in the A phase (Reproduced from (**a**) Ref. [11] and (**b**)–(**d**) Ref. [12])

Here, R_0 and S_A are coupling coefficients and P $(0 < P < 1)$ is the spin polarisation ratio of conduction electrons. B_{rz}^{em} is given by

$$B_{rz}^{em} = \Phi_0 \Phi^z, \tag{3.30}$$

with a single emergent magnetic flux quantum $\Phi_0 = h/e$ and skyrmion density Φ^z defined as

$$\Phi^\mu = \frac{1}{8\pi}\epsilon_{\mu\nu\lambda}\mathbf{m}\cdot(\partial_\nu\mathbf{m}\times\partial_\lambda\mathbf{m}). \tag{3.31}$$

Here, $\epsilon_{\mu\nu\lambda}$ is the totally anti-symmetric tensor and $\mathbf{m} = \mathbf{M}/|M|$. This means that each skyrmion induces exactly one quantum of emergent magnetic flux and that the resultant topological Hall resistivity ρ_{xy}^T is proportional to skyrmion density Φ^z. Figure 3.3d indicates the contribution of ρ_{xy}^T for MnSi, estimated by subtracting the B-linear term in ρ_{xy}. Consistent with the negative skyrmion number $\int dxdy\Phi^z = -1$ in the A phase (i.e. the skyrmion crystal state) and $\int dxdy\Phi^z = 0$ in the helical magnetic phase, the finite ρ_{xy}^T of the sign opposite from ρ_{xy}^N is observed only in the former spin state.

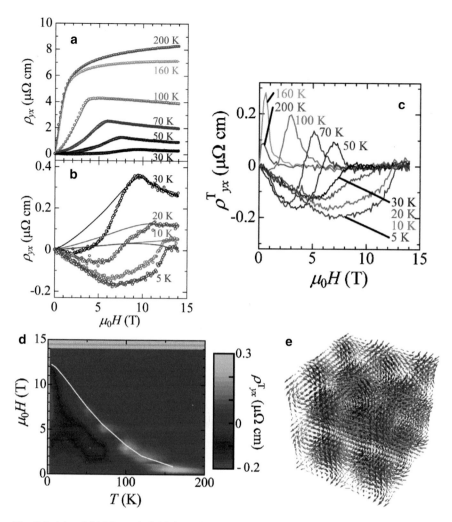

Fig. 3.4 (**a**) and (**b**) Magnetic field dependences of the Hall resistivity ρ_{yx} at various temperatures for MnGe. *Solid lines* are the fitting curves of ρ_{yx} using the relation $\rho_{yx} = R_0 B + S_A \rho_{xx} M$ with fitting parameters R_0 and S_A. (**c**) Magnetic field dependence of the topological Hall resistivity ρ_{yx}^T. (**d**) A contour map of ρ_{yx}^T in the plane of temperature and magnetic field. The *white curve* represents the temperature variation of the critical field H_c, at which a ferromagnetic spin-collinear state is realised. (**e**) Schematic of a cubic skyrmion lattice proposed as the magnetic ground state of MnGe (Reproduced from (**a**)–(**d**) Ref. [15] and (**e**) Ref. [16])

Note that Eq. (3.30) predicts that a higher skyrmion density (i.e. a smaller skyrmion size) gives larger values of B_{rz}^{em} and ρ_{xy}^T. Compared with $\rho_{xy}^T \sim 0.004\,\mu\Omega$ cm for MnSi with a spin modulation period $\lambda_m \sim 17$ nm, a 40-times-larger $\rho_{xy}^T \sim 0.16\,\mu\Omega$ cm has been reported for MnGe with $\lambda_m \sim 3$ nm (Fig. 3.4c). For MnGe, the topological Hall effect is observed in a much wider H–T range than the case for other B20 compounds characterised by a narrow magnetic A phase with a triangular lattice of skyrmions (Fig. 3.4d) [15]. Combined with a recent analysis by neutron-scattering experiments, the possible emergence of a unique cubic or square lattice of skyrmions has been proposed for MnGe (Fig. 3.4e) [16].

Interestingly, the topological Hall effect originating from a scalar spin chirality has also been reported for several material systems without a long-range magnetic order, such as pyrochlore $Pr_2Ir_2O_7$ with a "chiral spin liquid" state [17] and MnSi under high pressure [20]. When hydrostatic pressure is applied to MnSi, the helical magnetic order is gradually suppressed and finally disappears above the critical pressure $p_c \sim 14.6$ kbar (Fig. 3.5a). Despite the absence of a long-range magnetic order above p_c, neutron diffraction experiments have revealed the existence of magnetic Bragg scattering intensity everywhere on the surface of a sphere with radius $|q| \sim 0.043$ Å in the reciprocal space below 10 K [18]. In this high-pressure state, several anomalous transport properties such as non-Fermi liquid behaviour ($\rho_{xx} \propto T^{1.5}$) [19] and the topological Hall effect have been reported [20]. Figure 3.5b, c respectively show the T–H phase diagram and the corresponding development of ρ_{xy} taken at various pressures. The continuous evolution of the topological Hall resistivity from the A phase to the non-Fermi liquid state implies that topological characteristics of a skyrmion lattice remain in the latter exotic, non-Fermi liquid state even without a long-range magnetic order.

3.4   Manipulation by Electric Current

Next, we discuss the current-induced dynamics of skyrmions in metallic materials. Conduction electrons with finite spin polarisation can interact with local magnetic moments through the exchange process of their spin angular momentum and provide the effective torque on a local spin object such as a ferromagnetic domain wall [4, 21–23]. This process is called spin-transfer torque and is now widely used to realise the current-induced reversal of magnetisation for commercially available MRAM devices [24]. Here, the magnetic domain wall is driven parallel to the current direction and the threshold current density required to de-pin the domain wall is around 10^{11} A/m^2.

Similarly, skyrmions should be dragged along the current direction through the spin-transfer torque. As discussed in Sect. 3.2, however, the topological winding in a skyrmion spin texture causes an additional Magnus force, which drives skyrmions along the direction normal to external current. These phenomena were first investigated by Jonietz et al. through neutron diffraction experiments [25]. For the skyrmion crystal state of bulk MnSi, they simultaneously applied a temperature gradient and electric current along the same direction normal to an external magnetic field (Fig. 3.6g). This caused a spatial gradient of the Magnus force and induced the rotation of a skyrmion lattice observed as a rotational shift of magnetic Bragg reflections in the reciprocal space (Fig. 3.6). Later, Yu et al. successfully observed the current-induced translational motion of skyrmions in real space for FeGe by

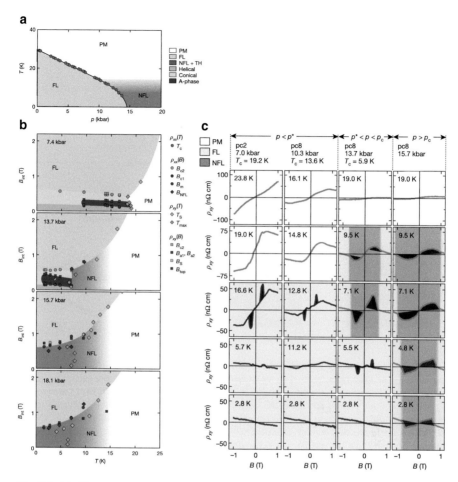

Fig. 3.5 (**a**) Temperature–pressure phase diagram under zero magnetic field for MnSi. (**b**) Temperature–magnetic field phase diagrams for MnSi with various magnitudes of hydrostatic pressure. Here FL, NFL and PM denote the Fermi liquid state, the non-Fermi liquid state and the paramagnetic state, respectively. (**c**) Magnetic field dependence of Hall resistivity measured at various temperatures and pressures (Reproduced from Ref. [20])

employing LTEM (Fig. 3.7) [26]. Notably, the threshold current density to drive skyrmions was $\sim 10^6$ A/m^2, which is five orders of magnitude smaller than that for typical magnetic domain walls (Fig. 3.6h and Fig. 3.7b). Such an ultra-low threshold current density is partly due to the particle nature of skyrmions, which allows them to effectively avoid pinning caused by impurity or defects, as discussed in Sect. 3.2.

According to Faraday's law of induction, the time development of magnetic flux induces electromotive force. By analogy, the generalisation of Faraday's law predicts that the time-dependent change of the Berry phase, i.e. the emergent magnetic flux, causes the emergent electric field $\mathbf{E}^{em}$, as discussed in Sect. 3.1.

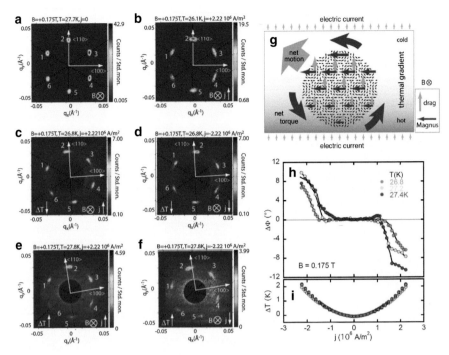

Fig. 3.6 Typical scattering intensity patterns for MnSi observed in neutron-scattering measurements with a neutron beam parallel to an applied magnetic field. The *red lines* serve as a guide to the eye. (**a**) Pattern for a skyrmion crystal in the absence of electric current. (**b**) Pattern in the presence of electric current flowing in the vertical direction (*arrow*). (**c**) When both current and a small anti-parallel temperature gradient are present, the scattering pattern rotates anti-clockwise. (**d**) Pattern when reversing current direction in (**c**). Those for reversed direction of the temperature gradient are shown in (**e**) and (**f**). (**g**) Schematic of spin-transfer-torque effects on a skyrmion lattice. A temperature gradient induces inhomogeneous Magnus and drag forces and therefore a rotational torque. (**h**) Change of the azimuthal angle of the rotation of the scattering pattern as a function of current density for three different temperatures (Reproduced from Ref. [25])

This emergent electric field is often called the spin-motive force [29], and its experimental detection has been reported for moving ferromagnetic domain walls [30] as well as an oscillating magnetic vortex in a ferromagnetic nanodisk [31]. Since the skyrmion spin texture is characterised by a quantised emergent magnetic flux, a skyrmion moving with velocity $\mathbf{v}_s$ induces an emergent electric field $\mathbf{E}^{em} = -\mathbf{v}_s \times \mathbf{B}_r^{em}$. When this skyrmion motion is driven by external current, the induced $\mathbf{E}^{em}$ partially cancels the electrical voltage induced by the topological Hall effect (Fig. 3.8a). Figure 3.8b shows the temperature dependence of ρ_{xy} measured by an alternating current lock-in method under the application of various magnitudes of additional direct current bias j [28]. The magnitude of ρ_{xy} in the skyrmion crystal

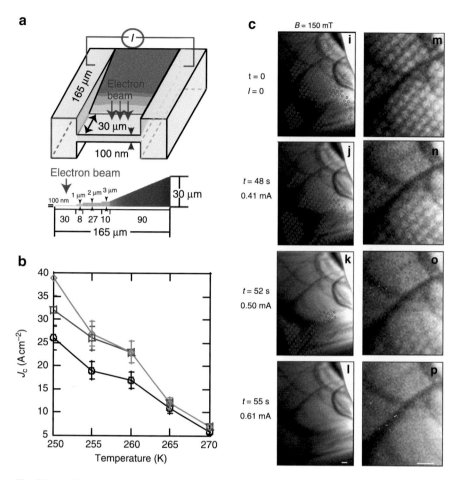

Fig. 3.7 (**a**) Schematic and cross-sectional view of a micro-device with a trapezoidal FeGe plate composed of a 100-nm-thick thinner terrace for electron beam transmission and another trapezoidal thicker part for supporting the thinner part. (**b**) Temperature dependence of critical current densities for skyrmion motion. (**c**) LTEM images of variations of a skyrmion crystal in slowly increasing current, as indicated by elapsed time and current values. The *right panel* indicates a magnified view of the *left panel*. Skyrmion positions are marked by red circles for clarity (Reproduced from Ref. [26])

state shows a sudden drop above the critical current density $j_c \sim 4 \times 10^5 \, \text{A/m}^2$ (Fig. 3.8c). This j_c value agrees well with the threshold current density to drive a translational skyrmion motion reported by the LTEM observation, suggesting that the observed drop of the ρ_{xy}-value above j_c indeed reflects the $\mathbf{E}^{\text{em}}$ induced by the skyrmion motion.

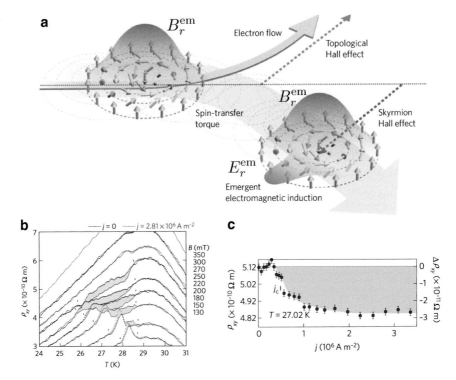

Fig. 3.8 (a) Schematic of skyrmion motion and associated physical phenomena under the flow of electrons. Electron current drives the flow of a skyrmion via the spin-transfer-torque mechanism. Electrons are deflected by the Lorentz force due to an emergent magnetic field $\mathbf{B}_r^{em}$ of the skyrmion, which results in the topological Hall effect. The velocity of the skyrmion has its transverse component, i.e. the skyrmion Hall effect. The motion of the skyrmion is accompanied by the time-dependent emergent magnetic field $\mathbf{B}_r^{em}$ and hence the emergent electric field $\mathbf{E}_r^{em}$. (b) The temperature dependence of the Hall resistivity ρ_{xy} in the skyrmion lattice phase of MnSi measured under various magnetic field magnitudes and applied d.c. electric currents. (c) Change of the Hall resistivity of MnSi as a function of the applied d.c. current at 250 mT. Above the critical current density j_c, the magnitude of ρ_{xy} suddenly decreases (Reproduced from (a) Ref. [27] and (b) and (c) Ref. [9])

Appendix 1: Landau-Lifshitz-Gilbert-Slonczewski Equation

The Landau–Lifshitz–Gilbert–Slonczewski equation containing the spin-transfer-torque term and non-adiabatic term (the so-called β term) introduced in Sect. 3.2 describes the magnetisation dynamics in the presence of spin-polarised electron currents. To analyse this equation numerically, we need to rewrite it in an appropriate form. In this Appendix, how to rewrite the equation in linearised, discretised and dimensionless forms is explained.

We start with the equation for local magnetisations $\mathbf{M}(\mathbf{r})$ $(=-\gamma\mathbf{S}(\mathbf{r}))$ as follows:

$$\frac{d\mathbf{M}}{dt} = -\gamma\mathbf{M} \times \left(-\frac{\partial\mathscr{H}}{\partial\mathbf{M}}\right) + \frac{\alpha}{M}\mathbf{M} \times \frac{d\mathbf{M}}{dt}$$

$$+\frac{pa^3}{2e}\frac{\gamma\hbar}{M}(\mathbf{j}\cdot\nabla)\mathbf{M} - \frac{pa^3\beta}{2e}\frac{\gamma\hbar}{M^2}[\mathbf{M}\times(\mathbf{j}\cdot\nabla)\mathbf{M}]. \tag{3.32}$$

Redefining the spatial gradient ∇ on a discretised square-lattice space, the third term is rewritten as

$$\frac{pa^2}{2e}\frac{\gamma\hbar}{M}\left(j_x\frac{\mathbf{M}_{i+\hat{\mathbf{x}}}-\mathbf{M}_{i-\hat{\mathbf{x}}}}{2} + j_y\frac{\mathbf{M}_{i+\hat{\mathbf{y}}}-\mathbf{M}_{i-\hat{\mathbf{y}}}}{2}\right) \equiv \mathscr{A}_i, \tag{3.33}$$

whereas the fourth term is written as

$$-\frac{pa^2\beta}{2e}\frac{\gamma\hbar}{M^2}\left[\mathbf{M}_i\times\left(j_x\frac{\mathbf{M}_{i+\hat{\mathbf{x}}}-\mathbf{M}_{i-\hat{\mathbf{x}}}}{2} + j_y\frac{\mathbf{M}_{i+\hat{\mathbf{y}}}-\mathbf{M}_{i-\hat{\mathbf{y}}}}{2}\right)\right] \equiv \mathscr{B}_i. \tag{3.34}$$

Then, the equation in the discretised form is as follows:

$$\frac{d\mathbf{M}_i}{dt} = -\gamma\mathbf{M}_i \times \left(-\frac{\partial\mathscr{H}}{\partial\mathbf{M}_i}\right) + \frac{\alpha}{M}\mathbf{M}_i \times \frac{d\mathbf{M}_i}{dt} + \mathscr{A}_i + \mathscr{B}_i. \tag{3.35}$$

Substituting this equation into $\frac{d\mathbf{M}_i}{dt}$ on its right-hand side, we obtain the equation in the linearised form as

$$\frac{d\mathbf{M}_i}{dt} = \frac{1}{1+\alpha^2}\left\{-\gamma\mathbf{M}_i \times \left(-\frac{\partial\mathscr{H}}{\partial\mathbf{M}_i}\right) + \mathscr{A}_i + \mathscr{B}_i\right.$$

$$\left. + \frac{\alpha}{M}\mathbf{M}_i \times \left[-\gamma\mathbf{M}_i \times \left(-\frac{\partial\mathscr{H}}{\partial\mathbf{M}_i}\right) + \mathscr{A}_i + \mathscr{B}_i\right]\right\}. \tag{3.36}$$

In the course of the derivation, we used a relation

$$\mathbf{M}_i \times \left(\mathbf{M}_i \times \frac{d\mathbf{M}_i}{dt}\right) = -M^2\frac{d\mathbf{M}_i}{dt}, \tag{3.37}$$

which comes from a vector formula $\mathbf{A}\times(\mathbf{B}\times\mathbf{C}) = \mathbf{B}(\mathbf{A}\times\mathbf{C})-\mathbf{C}(\mathbf{A}\cdot\mathbf{B})$. Introducing dimensionless local magnetisation $\mathbf{m}_i = \mathbf{M}_i/\gamma\hbar$, the equation in the dimensionless form is derived as

$$\frac{d\mathbf{m}_i}{d\tau} = \frac{1}{1+\alpha^2}\left\{-\mathbf{m}_i \times \left(-\frac{\partial\tilde{\mathscr{H}}}{\partial\mathbf{m}_i}\right) + \left(\tilde{\mathscr{A}}_i + \tilde{\mathscr{B}}_i\right)\right.$$

$$\left. + \frac{\alpha}{m}\mathbf{m}_i \times \left[-\mathbf{m}_i \times \left(-\frac{\partial\tilde{\mathscr{H}}}{\partial\mathbf{m}_i}\right) + \left(\tilde{\mathscr{A}}_i + \tilde{\mathscr{B}}_i\right)\right]\right\}, \tag{3.38}$$

with

$$\tilde{\mathscr{A}}_i = \frac{pa^2\hbar}{2emJ}\left(j_x\frac{\mathbf{m}_{i+\hat{\mathbf{x}}} - \mathbf{m}_{i-\hat{\mathbf{x}}}}{2} + j_y\frac{\mathbf{m}_{i+\hat{\mathbf{y}}} - \mathbf{m}_{i-\hat{\mathbf{y}}}}{2}\right),\tag{3.39}$$

$$\tilde{\mathscr{B}}_i = -\frac{pa^2\hbar}{2emJ}\frac{\beta}{m}\left[\mathbf{m}_i\times\left(j_x\frac{\mathbf{m}_{i+\hat{\mathbf{x}}} - \mathbf{m}_{i-\hat{\mathbf{x}}}}{2} + j_y\frac{\mathbf{m}_{i+\hat{\mathbf{y}}} - \mathbf{m}_{i-\hat{\mathbf{y}}}}{2}\right)\right],\tag{3.40}$$

where $\tau \equiv tJ/\hbar$ is the dimensionless time and $\tilde{\mathscr{H}} \equiv \mathscr{H}/J$ is the dimensionless Hamiltonian.

To transform the time unit, we need to multiply the dimensionless simulation time by the factor $\hbar/J$. For example, $\hbar/J \sim 6.6 \times 10^{-13}$ s for $J = 1$ meV. On the other hand, the dimension of the constant $\mathscr{K} = \frac{pa^2\hbar}{2emJ}$, which appears in $\tilde{\mathscr{A}}_i$ and $\tilde{\mathscr{B}}_i$, is $[\mathrm{A}^{-1}\cdot\mathrm{m}^2]$. For example, $\mathscr{K} = 0.82 \times 10^{-13}[\mathrm{A}^{-1}\cdot\mathrm{m}^2]$ when $J = 1$ meV $a = 5$ Å $= 5\times10^{-10}$ m $p = 0.2$ and $m = 1$. Therefore, $\mathscr{K}j$ becomes dimensionless, where $j[\mathrm{A}\cdot\mathrm{m}^{-2}]$ is the electric current density.

Appendix 2: Derivation of Thiele's Equation

Thiele's equation describes the centre-of-mass motion of a spin texture such as a skyrmion or a vortex under the assumption that its shape is rigid and never changes during the motion. The equation is derived from the Landau–Lifshitz–Gilbert–Slonczewski equation. In this Appendix, we derive Thiele's equation for a system with spin-polarised electric currents.

The Landau–Lifshitz–Gilbert–Slonczewski equation for the dynamics of local magnetisations $\mathbf{M}(\mathbf{r})$ $(=-\gamma\mathbf{S}(\mathbf{r}))$ is given by

$$\frac{d\mathbf{M}}{dt} = -\gamma\mathbf{M}\times\left(-\frac{\delta U}{\delta\mathbf{M}}\right) + \frac{\alpha_G}{M}\mathbf{M}\times\frac{d\mathbf{M}}{dt}$$
$$+ \frac{pa^3\gamma\hbar}{2eM}(\mathbf{j}\cdot\nabla)\mathbf{M} - \frac{pa^3\gamma\hbar}{2eM}\frac{\beta}{M}[\mathbf{M}\times(\mathbf{j}\cdot\nabla)\mathbf{M}],\tag{3.41}$$

where $\gamma = \frac{g\mu_B}{\hbar}(> 0)$ is the gyromagnetic ratio, $e(> 0)$ is the elementary charge, a is the lattice constant and α_G is the Gilbert damping coefficient. Here, the spin polarisation p represents the extent to which electric current is spin polarised as

$$p = \frac{\mathbf{j}_\uparrow - \mathbf{j}_\downarrow}{\mathbf{j}_\uparrow + \mathbf{j}_\downarrow} = \frac{\mathbf{j}_\uparrow - \mathbf{j}_\downarrow}{\mathbf{j}},\tag{3.42}$$

with $j_\uparrow$ and $j_\downarrow$ being the electric current density for up-magnetised and down-magnetised electrons (or equivalently for spin-down and spin-up electrons), respectively.

Let us consider the factor in the third and fourth terms:

$$-\frac{pa^3\gamma\hbar}{2eM}\mathbf{j}. \tag{3.43}$$

This factor turns out to be a velocity of conduction electrons multiplied by a certain factor. The current density $\mathbf{j}$ is given by

$$\mathbf{j} = -e\rho\mathbf{v_e}, \tag{3.44}$$

where ρ is the electron number density and $\mathbf{v_e}$ is the averaged electron velocity. The contribution to $\mathbf{j}$ from the up-magnetised electrons is written as

$$\mathbf{j_\uparrow} = -e\rho_\uparrow\mathbf{v_e}, \tag{3.45}$$

whereas that from the down-magnetised electrons is written as

$$\mathbf{j_\downarrow} = -e\rho_\downarrow\mathbf{v_e}. \tag{3.46}$$

According to the definition of spin polarisation p, we obtain

$$p\mathbf{j} = \mathbf{j_\uparrow} - \mathbf{j_\downarrow} = -e(\rho_\uparrow - \rho_\downarrow)\mathbf{v_e}. \tag{3.47}$$

Because each electron has a magnetisation of $m = \gamma\hbar/2 = 1\mu_B$ ($s = \hbar/2$ spin), the magnetisation current density $\mathbf{j_m}$ is given by

$$\mathbf{j_m} = -\frac{\gamma\hbar}{2e}p\mathbf{j} = -\frac{\gamma\hbar}{2e}(\mathbf{j_\uparrow} - \mathbf{j_\downarrow}) = \frac{\gamma\hbar}{2}(\rho_\uparrow - \rho_\downarrow)\mathbf{v_e}. \tag{3.48}$$

A local magnetisation $\mathbf{M(r)}$ on each site occupies a spatial volume of a^3. Therefore, the total conduction-electron magnetisations acting on $\mathbf{M(r)}$ per unit time is given by

$$-\frac{pa^3\gamma\hbar}{2e}\mathbf{j} = \frac{\gamma\hbar}{2}a^3(\rho_\uparrow - \rho_\downarrow)\mathbf{v_e} = \frac{\gamma\hbar}{2}(n_\uparrow - n_\downarrow)\mathbf{v_e} \equiv M_e\mathbf{v_e}, \tag{3.49}$$

where $n_\uparrow$ ($n_\downarrow$) is the number of up-magnetised (down-magnetised) electrons in the spatial volume a^3 and the quantity $\frac{\gamma\hbar}{2}(n_\uparrow - n_\downarrow)$ represents the total conduction-electron magnetisation M_e in the volume a^3. Dividing this equation by $M = |\mathbf{M}|$, we obtain

$$-\frac{pa^3\gamma\hbar}{2eM}\mathbf{j} = \frac{M_\mathrm{e}}{M}\mathbf{v}_\mathrm{e} \equiv \mathbf{v}_\mathrm{s}. \tag{3.50}$$

This equation indicates that the factor $-\frac{pa^3\gamma\hbar}{2eM}\mathbf{j}$ is the averaged velocity of conduction electrons multiplied by a factor M_e/M.

Using this notation, the Landau–Lifshitz–Gilbert–Slonczewski equation is rewritten as

$$\frac{d\mathbf{M}}{dt} = -\gamma\mathbf{M} \times \left(-\frac{\delta U}{\delta\mathbf{M}}\right) + \frac{\alpha_\mathrm{G}}{M}\mathbf{M} \times \frac{d\mathbf{M}}{dt}$$

$$-(\mathbf{v}_\mathrm{s} \cdot \nabla)\mathbf{M} + \frac{\beta}{M}\left[\mathbf{M} \times (\mathbf{v}_\mathrm{s} \cdot \nabla)\mathbf{M}\right]. \tag{3.51}$$

Dividing both sides of this equation by M and introducing $\mathbf{n} = \mathbf{M}/M$, we obtain

$$\dot{\mathbf{n}} = \mathbf{n} \times \frac{\gamma}{M}\left(\frac{\delta U}{\delta\mathbf{n}}\right) + \alpha_\mathrm{G}\mathbf{n} \times \dot{\mathbf{n}} - (\mathbf{v}_\mathrm{s} \cdot \nabla)\mathbf{n} + \beta\left[\mathbf{n} \times (\mathbf{v}_\mathrm{s} \cdot \nabla)\mathbf{n}\right], \tag{3.52}$$

where $\dot{\mathbf{n}} \equiv d\mathbf{n}/dt$. After multiplying by the vector $\mathbf{n}$, the equation reads

$$\mathbf{n} \times \dot{\mathbf{n}} = \mathbf{n} \times \left[\mathbf{n} \times \frac{\gamma}{M}\left(\frac{\delta U}{\delta\mathbf{n}}\right)\right] + \alpha_\mathrm{G}\mathbf{n} \times (\mathbf{n} \times \dot{\mathbf{n}})$$

$$-\mathbf{n} \times (\mathbf{v}_\mathrm{s} \cdot \nabla)\mathbf{n} + \beta\mathbf{n} \times \left[\mathbf{n} \times (\mathbf{v}_\mathrm{s} \cdot \nabla)\mathbf{n}\right]. \tag{3.53}$$

Using the vector formula $\mathbf{A} \times (\mathbf{B} \times \mathbf{C}) = (\mathbf{A} \cdot \mathbf{C})\mathbf{B} - (\mathbf{A} \cdot \mathbf{B})\mathbf{C}$, the first term of the right-hand side reads

$$\mathbf{n} \times \left[\mathbf{n} \times \frac{\gamma}{M}\left(\frac{\delta U}{\delta\mathbf{n}}\right)\right] = \left[\mathbf{n} \cdot \frac{\gamma}{M}\left(\frac{\delta U}{\delta\mathbf{n}}\right)\right]\mathbf{n} - \frac{\gamma}{M}\frac{\delta U}{\delta\mathbf{n}}, \tag{3.54}$$

whereas the second term reads

$$\alpha_\mathrm{G}\mathbf{n} \times (\mathbf{n} \times \dot{\mathbf{n}}) = \alpha_\mathrm{G}\left[(\mathbf{n} \cdot \dot{\mathbf{n}})\mathbf{n} - (\mathbf{n} \cdot \mathbf{n})\dot{\mathbf{n}}\right] = -\alpha_\mathrm{G}\dot{\mathbf{n}}. \tag{3.55}$$

Here, we used the relations $\mathbf{n} \cdot \mathbf{n} = 1$ and $\mathbf{n} \cdot \dot{\mathbf{n}} = 0$. The latter relation holds because the vector $\mathbf{n}$ has a constant norm ($= 1$), and thus, the vector $\dot{\mathbf{n}}$ has only a component normal to the radial direction. The fourth term reads

$$\beta\mathbf{n} \times \left[\mathbf{n} \times (\mathbf{v}_\mathrm{s} \cdot \nabla)\mathbf{n}\right] = \beta\left[\mathbf{n} \cdot (\mathbf{v}_\mathrm{s} \cdot \nabla)\mathbf{n}\right]\mathbf{n} - \beta(\mathbf{v}_\mathrm{s} \cdot \nabla)\mathbf{n}. \tag{3.56}$$

Note that we used the above vector formula again.

At present, we have

$$\mathbf{n} \times \dot{\mathbf{n}} = \left[\mathbf{n} \cdot \frac{\gamma}{M} \left(\frac{\delta U}{\delta \mathbf{n}} \right) \right] \mathbf{n} - \frac{\gamma}{M} \frac{\delta U}{\delta \mathbf{n}} - \alpha_G \dot{\mathbf{n}}$$

$$-\mathbf{n} \times (\mathbf{v}_s \cdot \nabla) \mathbf{n} + \beta \left[\mathbf{n} \cdot (\mathbf{v}_s \cdot \nabla) \mathbf{n} \right] \mathbf{n} - \beta (\mathbf{v}_s \cdot \nabla) \mathbf{n}. \qquad (3.57)$$

Now, we introduce the assumption that the magnetic structure is never deformed during its motion. That is, we assume that a function $\mathbf{n}(\mathbf{r}, t)$, which describes a magnetisation direction at time t and position $\mathbf{r}$, depends only on relative coordinates $\boldsymbol{\xi}(t) \equiv \mathbf{r} - \mathbf{R}(t)$ measured from centre-of-mass coordinates $\mathbf{R}(t)$ as

$$\mathbf{n}(\mathbf{r}, t) = \mathbf{n}(\boldsymbol{\xi}(t)) = \mathbf{n}(\mathbf{r} - \mathbf{R}(t)). \qquad (3.58)$$

Under this assumption, the relation that holds is as follows:

$$\dot{\mathbf{n}}(\mathbf{r} - \mathbf{R}(t)) = \sum_j \dot{\xi}_j \frac{\partial \mathbf{n}}{\partial \xi_j} = -\sum_j \dot{R}_j \frac{\partial \mathbf{n}}{\partial \xi_j} = -\left(\dot{X} \frac{\partial \mathbf{n}}{\partial x} + \dot{Y} \frac{\partial \mathbf{n}}{\partial y} + \dot{Z} \frac{\partial \mathbf{n}}{\partial z} \right)$$

$$= -(\mathbf{v}_d \cdot \nabla) \mathbf{n}, \qquad (3.59)$$

where $\mathbf{R} = (R_1, R_2, R_3) = (X, Y, Z)$ are centre-of-mass coordinates of the magnetic structure, $\boldsymbol{\xi} = (\xi_1, \xi_2, \xi_3) = (x, y, z)$ are relative coordinates and $\dot{\mathbf{R}} = (\dot{X}, \dot{Y}, \dot{Z}) \equiv \mathbf{v}_d$ is the velocity of the centre-of-mass motion. Hereafter, we omit the summation sign $\sum_j$.

Substituting this relation into Eq. (3.59), we obtain

$$\mathbf{n} \times [(\mathbf{v}_s - \mathbf{v}_d) \cdot \nabla] \mathbf{n} + [(\beta \mathbf{v}_s - \alpha_G \mathbf{v}_d) \cdot \nabla] \mathbf{n} + \frac{\gamma}{M} \frac{\delta U}{\delta \mathbf{n}}$$

$$= \left[\mathbf{n} \cdot \frac{\gamma}{M} \left(\frac{\delta U}{\delta \mathbf{n}} \right) \right] \mathbf{n} + \beta \left[\mathbf{n} \cdot (\mathbf{v}_s \cdot \nabla) \mathbf{n} \right] \mathbf{n}. \qquad (3.60)$$

Taking the inner product with $-\partial \mathbf{n} / \partial \xi_i$, we get

$$-\frac{\partial \mathbf{n}}{\partial \xi_i} \cdot \left[\mathbf{n} \times (v_s - v_d)_j \frac{\partial \mathbf{n}}{\partial \xi_j} \right] - \frac{\partial \mathbf{n}}{\partial \xi_i} \cdot (\beta v_s - \alpha_G v_d)_j \frac{\partial \mathbf{n}}{\partial \xi_j} - \frac{\gamma}{M} \frac{\partial \mathbf{n}}{\partial \xi_i} \cdot \frac{\delta U}{\delta \mathbf{n}} = 0.$$

$$(3.61)$$

Note that the relation $\frac{\partial \mathbf{n}}{\partial \xi_i} \cdot \mathbf{n} = 0$ holds because a spatial variation $\frac{\partial \mathbf{n}}{\partial \xi_i}$ always occurs a in the direction normal to the vector $\mathbf{n}$.

Using the vector formula $\mathbf{A} \times (\mathbf{B} \times \mathbf{C}) = (\mathbf{A} \cdot \mathbf{C})\mathbf{B} - (\mathbf{A} \cdot \mathbf{B})\mathbf{C}$, we obtain

$$\mathbf{n} \cdot \left(\frac{\partial \mathbf{n}}{\partial \xi_i} \times \frac{\partial \mathbf{n}}{\partial \xi_j} \right) (v_s - v_d)_j - \frac{\partial \mathbf{n}}{\partial \xi_i} \cdot \frac{\partial \mathbf{n}}{\partial \xi_j} (\beta v_s - \alpha_G v_d)_j - \frac{\gamma}{M} \frac{\partial \mathbf{n}}{\partial \xi_i} \cdot \frac{\delta U}{\delta \mathbf{n}} = 0. \tag{3.62}$$

After integrating both sides of the equation in three-dimensional space and multiplying by M/γ, the equation reads

$$\frac{M}{\gamma} (v_s - v_d)_j \int \mathbf{n} \cdot \left(\frac{\partial \mathbf{n}}{\partial \xi_i} \times \frac{\partial \mathbf{n}}{\partial \xi_j} \right) d^3\mathbf{r} - \frac{M}{\gamma} (\beta v_s - \alpha_G v_d)_j \int \frac{\partial \mathbf{n}}{\partial \xi_i} \cdot \frac{\partial \mathbf{n}}{\partial \xi_j} d^3\mathbf{r}$$

$$+ \nabla_R U = 0, \tag{3.63}$$

because the third term is calculated as

$$-\int \frac{\partial \mathbf{n}}{\partial \xi_i} \cdot \frac{\delta U}{\delta \mathbf{n}} d^3\mathbf{r} = -\frac{\partial U}{\partial \xi_i} = \frac{\partial U}{\partial R_i} = \nabla_R U. \tag{3.64}$$

When magnetisations stack ferromagnetically along the z direction in thin-plate specimens with a thickness of d, the integration with respect to z can be replaced by multiplication by the number of stacked layers d/a:

$$\frac{Md}{\gamma a} (v_s - v_d)_j \int \mathbf{n} \cdot \left(\frac{\partial \mathbf{n}}{\partial \xi_i} \times \frac{\partial \mathbf{n}}{\partial \xi_j} \right) dxdy - \frac{Md}{\gamma a} (\beta v_s - \alpha_G v_d)_j \int \frac{\partial \mathbf{n}}{\partial \xi_i} \cdot \frac{\partial \mathbf{n}}{\partial \xi_j} dxdy$$

$$+ \nabla_R U = 0. \tag{3.65}$$

This equation can be rewritten as

$$G_{ij} (v_s - v_d)_j - D_{ij} (\beta v_s - \alpha_G v_d)_j - F_i = 0, \tag{3.66}$$

where

$$G_{ij}(\boldsymbol{\xi}) = \frac{Md}{\gamma a} \int \mathbf{n} \cdot \left(\frac{\partial \mathbf{n}}{\partial \xi_i} \times \frac{\partial \mathbf{n}}{\partial \xi_j} \right) dxdy$$

$$= \begin{cases} \mathcal{G} & \text{for } (\xi_i, \xi_j) = (x, y) \\ -\mathcal{G} & \text{for } (\xi_i, \xi_j) = (y, x) \\ 0 & \text{otherwise} \end{cases} \tag{3.67}$$

$$D_{ij}(\boldsymbol{\xi}) = \frac{Md}{\gamma a} \int \frac{\partial \mathbf{n}}{\partial \xi_i} \cdot \frac{\partial \mathbf{n}}{\partial \xi_j} \, dxdy$$

$$= \begin{cases} \mathscr{D} & \text{for } (\xi_i, \xi_j) = (x, x), (y, y) \\ 0 & \text{otherwise} \end{cases}$$

$$(3.68)$$

Note that the factor $\frac{Md}{\gamma a}$ becomes unity for the two-dimensional system ($d/a = 1$) when the norm of the dimensionless magnetisation vector $\mathbf{M}/\gamma$ is taken to be unity. For topological spin textures such as skyrmion and vortices, $\mathscr{G}$ becomes finite.

The first term of the above equation can be rewritten as

$$G_{ij}(v_s - v_d)_j = \begin{cases} G_{xx}(v_s - v_d)_x + G_{xy}(v_s - v_d)_y + G_{xz}(v_s - v_d)_z & \text{for } i = x \\ G_{yx}(v_s - v_d)_x + G_{yy}(v_s - v_d)_y + G_{yz}(v_s - v_d)_z & \text{for } i = y \\ G_{zx}(v_s - v_d)_x + G_{zy}(v_s - v_d)_y + G_{zz}(v_s - v_d)_z & \text{for } i = z \end{cases}$$

$$= \begin{cases} \mathscr{G}(v_s - v_d)_y & \text{for } i = x \\ -\mathscr{G}(v_s - v_d)_x & \text{for } i = y \\ 0 & \text{for } i = z \end{cases}$$

$$= -\mathbf{G} \times (\mathbf{v}_s - \mathbf{v}_d), \qquad (3.69)$$

where $\mathbf{G} = (0, 0, \mathscr{G})$ is referred to as the gyromagnetic coupling vector.

According to Eq. (3.68), the third term can be rewritten as

$$-D_{ij}(v_s - v_d)_j = -\mathscr{D}(\mathbf{v}_s - \mathbf{v}_d). \qquad (3.70)$$

Eventually, the expression for Thiele's equation is obtained as

$$\mathbf{G} \times (\mathbf{v}_s - \mathbf{v}_d) + \mathscr{D}(\beta \mathbf{v}_s - \alpha_G \mathbf{v}_d) + \mathbf{F} = 0, \qquad (3.71)$$

or

$$\mathbf{G} \times (\mathbf{v}_s - \mathbf{v}_d) + \mathscr{D}(\beta \mathbf{v}_s - \alpha_G \mathbf{v}_d) - \nabla U = 0, \qquad (3.72)$$

where $\mathbf{G} = (0, 0, \mathscr{G})$. In the latter equation, the sign $\nabla_R = \frac{\partial}{\partial R_i} = (\frac{\partial}{\partial X}, \frac{\partial}{\partial Y}, \frac{\partial}{\partial Z})$ is simply rewritten as ∇.

For a skyrmion, $\mathscr{G}$ is given by $\mathscr{G} = 4\pi nQ \cdot \frac{Md}{\gamma a}$, where the winding number n expresses the number of times magnetisations wrap a sphere and $Q(= \pm 1)$ represents the direction of magnetisation at the skyrmion core, i.e. $Q = +1$ for up magnetisation and $Q = -1$ for down magnetisation. On the other hand, $\mathscr{G}$ is $\mathscr{G} = 2\pi Q \cdot \frac{Md}{\gamma a}$ for a magnetic vortex, whereas $\mathscr{G} = 0$ for helices and ferromagnetic domain walls.

References

1. G. Volovik, J. Phys. C **20**, L87 (1987)
2. N. Nagaosa, Y. Tokura, Phys. Scr. **T146**, 014020 (2012)
3. N. Nagaosa, X.Z. Yu, Y. Tokura, Phil. Trans. R. Soc. A **370**, 5806 (2012)
4. J.C. Slonczewski, J. Magn. Magn. Mater. **159**, L1 (1996)
5. J. Shibata, Y. Nakatani, G. Tatara, H. Kohno, Y. Ohtani, Phys. Rev. B **73**, 020403(R) (2006)
6. J. Iwasaki, M. Mochizuki, N. Nagaosa, Nat. Commun. **4**, 1463 (2013)
7. A.A. Thiele, Phys. Rev. Lett. **30**, 230 (1973)
8. K. Everschor, M. Garst, B. Binz, F. Jonietz, S. Mühlbauer, C. Pfleiderer, A. Rosch, Phys. Rev. B **86**, 054432 (2012)
9. T. Schulz, R. Ritz, A. Bauer, M. Halder, M. Wagner, C. Franz, C. Pfleiderer, K. Everschor, M. Garst, A. Rosch, Nat. Phys. **8**, 301 (2012)
10. J. Iwasaki, M. Mochizuki, N. Nagaosa, Nat. Nanotech. **8**, 742 (2013)
11. C. Pfleiderer, A. Rosch, Nature **465**, 880 (2010)
12. A. Neubauer, C. Pfleiderer, B. Binz, A. Rosch, R. Ritz, P.G. Niklowitz, P. Böni, Phys. Rev. Lett. **102**, 186602 (2009)
13. M. Onoda, G. Tatara, N. Nagaosa, J. Phys. Soc. Jpn. **73**, 2624 (2004)
14. Y. Taguchi, Y. Oohara, H. Yoshizawa, N. Nagaosa, Science **291**, 2573 (2001)
15. N. Kanazawa, Y. Onose, T. Arima, D. Okuyama, K. Ohoyama, S. Wakimoto, K. Kakurai, S. Ishiwata, Y. Tokura, Phys. Rev. Lett. **106**, 156603 (2011)
16. N. Kanazawa, J.-H. Kim, D.S. Inosov, J.S. White, N. Egetenmeyer, J.L. Gavilano, S. Ishiwata, Y. Onose, T. Arima, B. Keimer, Y. Tokura, Phys. Rev. B **86**, 134425 (2012)
17. Y. Machida, S. Nakatsuji, S. Onoda, T. Tayama, T. Sakakibara, Nature **463**, 210 (2010)
18. C. Pfleiderer, D. Reznik, L. Pintschovius, H.v. Löhneysen, M. Garst, A. Rosch, Nature **427**, 227 (2004)
19. C. Pfleiderer, P. Böni, T. Keller, U.K. Rössler, A. Rosch, Science **316**, 1871 (2007)
20. R. Ritz, M. Halder, M. Wagner, C. Franz, A. Bauer, C. Pfleiderer, Nature **497**, 231 (2013)
21. L. Berger, Phys. Rev. B **54**, 9353 (1996)
22. J. Grollier, P. Boulenc, V. Cros, A. Hamzić, Vaurés, A. Fert, G. Faini, Appl. Phys. Lett. **83**, 509 (2003)
23. M. Tsoi, R. Fontana, S. Parkin, Appl. Phys. Lett. **83**, 2617 (2003)
24. A.V. Khvalkovskiy, D. Apalkov, S. Watts, R. Chepulskii, R.S. Beach, A. Ong, X. Tang, A. Driskill-Smith, W.H. Butler, P.B. Visscher, D. Lottis, E. Chen, V. Nikitin, M. Krounbi, J. Phys. D Appl. Phys. **46**, 074001 (2013)
25. F. Jonietz, S. Mühlbauer, C. Pfleiderer, A. Neubauer, W. Münzer, A. Bauer, T. Adams, R. Georgii, P. Böni, R.A. Duine, K. Everschor, M. Garst, A. Rosch, Science **330**, 1648 (2010)
26. X.Z. Yu, N. Kanazawa, W.Z. Zhang, T. Nagai, T. Hara, K. Kimoto, Y. Matsui, Y. Onose, Y. Tokura, Nat. Commun. **3**, 988 (2012)
27. N. Nagaosa, Y. Tokura, Nat. Nanotech. **8**, 899 (2013)
28. T. Schulz, R. Ritz, A. Bauer, M. Halder, M. Wagner, C. Franz, C. Pfleiderer, K. Everschor, M. Garst, A. Rosch, Nat. Phys. **8**, 301 (2012)
29. S.E. Barnes, S. Maekawa, Phys. Rev. Lett. **98**, 246601 (2007)
30. M. Hayashi, J. Ieda, Y. Yamane, J. Ohe, Y.K. Takahashi, S. Mitani, S. Maekawa, Phys. Rev. Lett. **108**, 147202 (2012)
31. K. Tanabe, D. Chiba, J. Ohe, S. Kasai, H. Kohno, S.E. Barnes, S. Maekawa, K. Kobayashi, T. Ono, Nat. Comm. **3** 845 (2012)

Chapter 4
Skyrmions and Electric Fields in Insulating Materials

Abstract In insulating materials, conduction electrons and associated emergent fields are absent; instead, magnetic skyrmions in insulators induce spatially inhomogeneous charge distributions through the relativistic spin-orbit interaction. Depending on the symmetry of an underlying crystallographic lattice, skyrmions carry electric dipoles or quadrupoles and can be manipulated by an external electric field. This property may provide an energetically more efficient method to control skyrmions because the electric field in an insulating system causes only negligible Joule heat loss compared with the current-driven approach in a metallic system. In this chapter, this magnetoelectric nature of skyrmions is discussed. Skyrmions also show resonant oscillation against both ac magnetic and electric fields of gigahertz frequency. The interference of these excitations causes unique optical responses called directional dichroism, where the sign reversal of light (microwave) propagation direction gives different absorption spectra.

4.1 Magnetoelectric Skyrmions and Manipulation by Electric Fields

The control of magnetism by electric fields E and of dielectric properties by magnetic fields H are called magnetoelectric (ME) effects and have long been studied since their first prediction by P. Curie more than 100 years ago [1]. When both time reversal and space-inversion symmetries are simultaneously broken in a material, the emergence of linear ME effects ($P_i = \alpha_{ij}H_j$ and $M_i = \alpha_{ji}E_j$, with P representing the macroscopic electric polarisations) can be allowed. While the magnitude of the linear ME coefficient α_{ij} is generally very small, the employment of materials with strongly coupled magnetic and dielectric orders (i.e. multiferroic materials) can lead to more versatile and gigantic ME responses [2, 3]. For example, in some frustrated magnets such as perovskite $TbMnO_3$, the emergence of helical spin order is found to induce ferroelectric polarisation [4]. Such spin-driven ferroelectricity originates from symmetry breaking by spin textures [5–7], and the utilisation of this strong ME coupling has enabled the H-induced reversal of P [8] and the E-induced reversal of M [9].

Similarly, the generalised concept of ME coupling can be also applied to skyrmion spin texture. Here, we discuss the case for the insulator Cu_2OSeO_3, whose

© Springer International Publishing Switzerland 2016
S. Seki, M. Mochizuki, *Skyrmions in Magnetic Materials*, SpringerBriefs in Physics, DOI 10.1007/978-3-319-24651-2_4

crystal lattice is characterised by the chiral cubic space group $P2_13$ (Fig. 2.1b). This compound contains two distinctive magnetic Cu^{2+} ($S = 1/2$) sites surrounded by different oxygen coordinations with a ratio of 3 : 1, leading to the stabilisation of three-up, one-down-type local ferrimagnetic ordering [10]. The magnetic phase diagram of Cu_2OSeO_3 is shown in Fig. 2.1d. While a helical spin order is realised in a ground state, the application of a finite magnetic field H stabilises a skyrmion lattice state in a narrow temperature region just below $T_c \sim 59$ K [11]. In Fig. 4.1m, the H dependence of electric polarisation P measured with $H \parallel [111]$ at 57 K is shown. For all ferrimagnetic, helimagnetic and skyrmion lattice spin states, the emergence of $P \parallel H \parallel [111]$ can be observed but with different signs and magnitudes. We can also observe $P \parallel [001]$ for $H \parallel [110]$ (Fig. 4.1i), while no electric polarisation can be detected for $H \parallel [001]$ (Fig. 4.1e) [12].

These behaviours can be well explained from the viewpoint of symmetry. The original crystal lattice of Cu_2OSeO_3 belongs to the non-polar space group $P2_13$ and does not have electric polarisation by itself (Fig. 4.1a). Similarly, the skyrmion lattice spin texture characterised by the orthogonal arrangement of the six-fold rotation axis parallel to H and $2'$ (two-fold rotation followed by time reversal) axes normal to H is also non-polar (Fig. 4.1b). However, when this skyrmion lattice spin texture is placed on the crystal lattice of Cu_2OSeO_3, most of the symmetry elements are broken and the system can become polar. In the case of $H \parallel [111]$, only a three-fold rotation axis parallel to H remains unbroken and the emergence of $P \parallel H \parallel [111]$ is allowed (Fig. 4.1n). For $H \parallel [110]$, only the screw axis normal to H survives and $P \parallel [001]$ emerges normal to H (Fig. 4.1j). In contrast, with $H \parallel [001]$, the orthogonal arrangement of three screw axes remains and the induction of P is prohibited (Fig. 4.1f). The above symmetry-based analyses well reproduce the experimental observations.

Because the magnetic modulation period (~ 62 nm) is much longer than the crystallographic lattice constant (~ 8.9 Å) for Cu_2OSeO_3, spin directions within a single crystallographic unit cell can be considered to be almost collinear. When a local magnetic moment $\mathbf{m}_i = (m_{ia}, m_{ib}, m_{ic})$ is assumed for the ith crystallographic unit cell, the $P2_13$ symmetry of the crystal lattice allows the emergence of a local electric polarisation $\mathbf{p}_i$ of the form

$$\mathbf{p}_i = (p_{ia}, p_{ib}, p_{ic}) \propto (m_{ib}m_{ic}, m_{ic}m_{ia}, m_{ia}m_{ib}), \qquad (4.1)$$

from Neumann's law and cluster expansion up to second order. In Fig. 4.2b, c, the expected relation between local directions of $\mathbf{m}_i$ and $\mathbf{p}_i$ is visualised. This local correspondence between $\mathbf{m}_i$ and $\mathbf{p}_i$ directly gives a macroscopic relation between M and P for a collinear ferrimagnetic state with uniformly oriented local magnetisation. To confirm the validity of Eq. (4.1), the H direction dependence of P is measured at 2 K for the collinear ferrimagnetic state (Fig. 4.2d, e). The experimental data agrees with the theoretical fit given by Eq. (4.1), which proves that this equation well describes the local ME coupling for Cu_2OSeO_3. Here, a microscopic origin of the ME coupling can be ascribed to the so-called spin-dependent d-p hybridisation mechanism [13, 14]. This model assumes a pair of

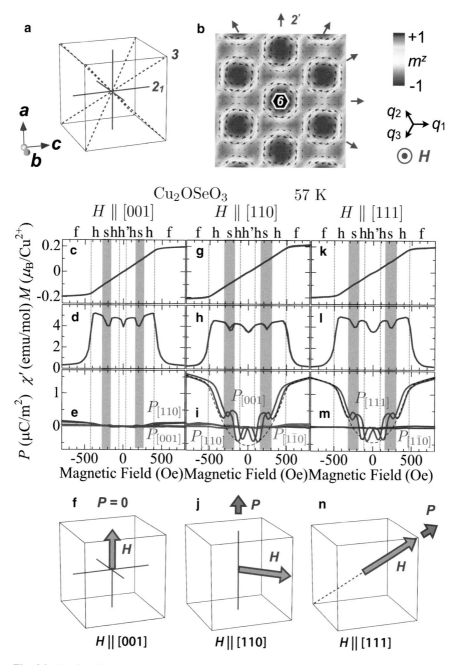

Fig. 4.1 (continued)

magnetic ions (accompanied by a magnetic moment $\mathbf{s}_i$) and a ligand ion (Fig. 4.2a). Because the direction of $\mathbf{s}_i$ affects the strength of covalence between these sites through the spin-orbit interaction, the local electric polarisation $\mathbf{p}_{ij}$ is induced along the bond direction. When the unit vector along the bond direction is defined as $\mathbf{e}_{ij}$, the relation

$$\mathbf{p}_{ij} \propto (\mathbf{e}_{ij} \cdot \mathbf{s}_i)^2 \mathbf{e}_{ij} \tag{4.2}$$

can be expected. By taking the summation $\sum_{ij} \mathbf{p}_{ij}$ over all Cu–O bonds within a crystallographic unit cell of Cu_2OSeO_3 assuming a collinear spin arrangement, we can reproduce Eq. (4.1). The validity of this ME-coupling mechanism for Cu_2OSeO_3 has also been supported by a recent calculation based on the density functional theory [15].

By identifying the manner of the local ME coupling, the spatial distribution of electric charge $\rho(\mathbf{r}) = -\nabla \cdot \mathbf{p}(\mathbf{r})$ for the given spin texture $\mathbf{m}(\mathbf{r})$ can be estimated using Eq. (4.1). In Fig. 4.3, electric charge distributions estimated for a skyrmion spin state with various directions of H are summarised. We can see that skyrmions under $H \parallel [110]$ ($H \parallel [001]$) carry local electric dipoles (quadrupoles), which strongly suggests that the translational motion of individual skyrmion particles can be driven by the spatial gradient of an external electric field. Recently, through a small-angle neutron diffraction experiment under the application of an electric field, the E-induced rotation of a skyrmion lattice has been reported [16, 17]. While a microscopic origin of the latter phenomena has not yet been completely resolved, the above results clearly demonstrate that the manipulation of skyrmions by electric fields in insulators is indeed possible. Because this approach is free from energy losses due to Joule heating in principle, the ME nature of skyrmions in insulators may contribute to further reduction of energy consumption associated with skyrmion manipulation.

Fig. 4.1 Symmetry elements compatible with (**a**) a crystal lattice of Cu_2OSeO_3 with chiral cubic space group $P2_13$ and (**b**) a magnetic skyrmion lattice formed within a plane normal to an applied magnetic field (H). In (**b**), the skyrmion lattice holds 6 (six-fold rotation) axes parallel to H and 2′ (two-fold rotation followed by time reversal) axes normal to H. (**c**)–(**e**) Magnetic field dependence of (**c**) magnetisation M, (**d**) ac magnetic susceptibility χ', and (**e**) electric polarisation P for bulk Cu_2OSeO_3 measured at 57 K with $H \parallel [001]$. Corresponding profiles for $H \parallel [110]$ and $H \parallel [111]$ are also indicated in (**g**)–(**i**) and (**k**)–(**m**), respectively. The letter symbols f, s, h and h' indicate the ferrimagnetic, skyrmion-crystal, helimagnetic (single q-domain) and helimagnetic (multiple q-domains) states, respectively. (**f**), (**j**), (**n**) Magnetically-induced ferroelectric polarisation (P) under various directions of H for Cu_2OSeO_3, predicted by symmetry analysis (see text) (Reproduced from Ref. [11] and Ref. [12])

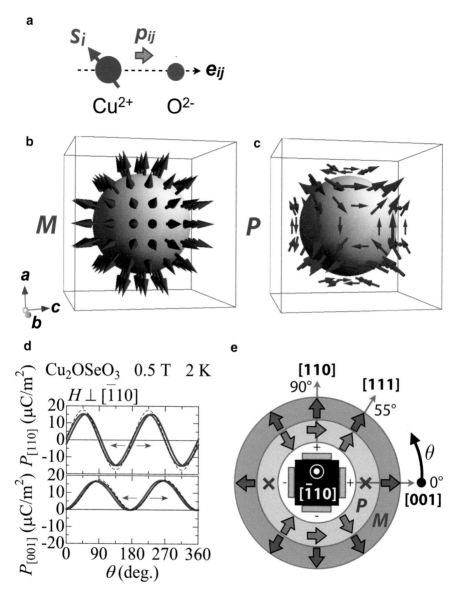

Fig. 4.2 (a) Schematic of the spin-dependent d-p hybridisation model. (b) and (c) Three-dimensional representation of the general correspondence between M and P directions in a collinear spin state for Cu_2OSeO_3, deduced based on Eq. (4.1). *Arrows* at the same position in (b) and (c) represent the M vector and corresponding induced P vector, respectively. (d) [110] and [001] components of electric polarisation P simultaneously measured under H rotating around the [$\bar{1}$10] axis at 2 K with $H = 0.5$ T (i.e. collinear ferrimagnetic state). *Dashed lines* indicate theoretically expected behaviours based on Eq. (4.1), and *arrows* denote the direction of H rotation. In (e), experimentally obtained relations between P and M directions in a ferrimagnetic state as well as the definition of θ (the angle between the H direction and [001] axis) are summarised. Here, M and P directions are indicated; the cross symbol $\times$ denotes $P = 0$ (Reproduced from Ref. [12])

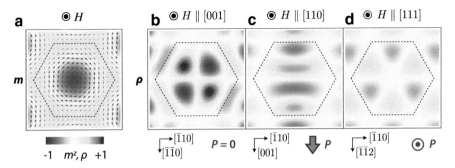

Fig. 4.3 Calculated spatial distributions of (**a**) the local magnetisation vector **m** and (**b**)–(**d**) the local electric charge ρ for the skyrmion lattice state (see text). The magnetic field is along the out-of-plane direction, and the results for (**b**) $H \parallel [001]$, (**c**) $H \parallel [110]$ and (**d**) $H \parallel [111]$ are indicated. The background colour represents the relative value of m_z for (**a**) and ρ for (**b**)–(**d**). Here, m_z indicates the out-of-plane component of **m**. The *dashed hexagon* indicates a magnetic unit cell of the skyrmion lattice or a single skyrmion quasi-particle (Reproduced from Ref. [12])

4.2 Magnetoelectric Resonance of Skyrmions

Thus far, the response of skyrmions to a static external field has mainly been discussed. Hereafter, we introduce the dynamics of skyrmions under oscillating magnetic and electric fields (H^ω and E^ω). In general, the appropriate frequency of H^ω induces the coherent resonance precession of magnetic moments in a long-range spin-ordered state. In conventional ferromagnets, such a magnetic resonance mode can be excited by H^ω applied normal to a static magnetic field H.

For the skyrmion lattice spin state, theoretical studies have predicted several magnetic resonance modes characterised by different selection rules [18, 19]. For the $H \perp H^\omega$ ($H \parallel H^\omega$) configuration, the clockwise or anticlockwise rotational modes (breathing modes) of skyrmions are expected (Fig. 4.4a, d). To experimentally identify these skyrmion resonance modes, a microwave absorption spectrum has been obtained for Cu_2OSeO_3 under various directions of H and H^ω [20]. Figure 4.4b indicates the H dependence of the absorption spectrum at 57.5 K (i.e. just below $T_c \sim 59$ K) with the $H \perp H^\omega$ setup. Here, the application of H induces the successive magnetic phase transitions (helical $\rightarrow$ skyrmion $\rightarrow$ helical $\rightarrow$ ferrimagnetic). While a helical spin state is always characterised by magnetic resonance at around $1.5 \sim 1.8$ GHz, the emergence of a new resonant mode at 1.0 GHz is clearly observed for the intermediate field region from 140 Oe to 320 Oe. The peak intensity of this new mode is plotted as a function of temperature and magnetic field in Fig. 4.4c, and we can see that it appears only in the skyrmion lattice spin state. On the basis of the previous theoretical prediction, this new mode is identified as the rotational mode of skyrmions. Moreover, a similar measurement has been performed for the $H \parallel H^\omega$ setup. The absorption peak by magnetic resonance is observed only in an intermediate field region from 50 Oe to 120 Oe (Fig. 4.4e) and is confirmed to appear only in the skyrmion lattice spin state (Fig. 4.4f). From its

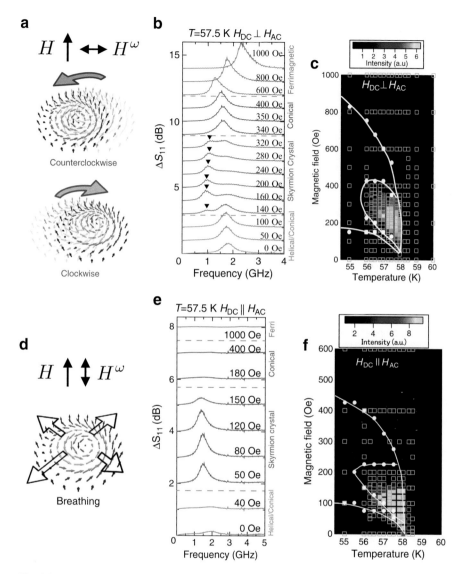

Fig. 4.4 (**a**) Clockwise and anticlockwise rotational modes and (**d**) breathing modes of magnetic skyrmions excited by $H \perp H^{\omega}$ and $H \parallel H^{\omega}$, respectively. The microwave absorption spectra under various magnitudes of a static magnetic field at 57.5 K for bulk Cu_2OSeO_3 ((**b**) and (**e**)), as well as the temperature versus magnetic field phase diagram with background colour indicating the absorption intensity of skyrmion resonant modes ((**c**) and (**f**)), are indicated for each experimental configuration of $H \perp H^{\omega}$ and $H \parallel H^{\omega}$, respectively (Reproduced from; (**a**) and (**d**) Ref. [27], (**b**), (**c**), (**e**), (**f**) Ref. [20])

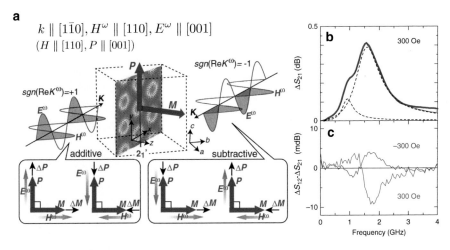

Fig. 4.5 (**a**) Configuration for microwave directional dichroism, where microwaves with positive and negative directions of a propagation vector (i.e. $+k$ and $-k$) show different absorption spectra. Here, the $H \parallel M \parallel [110]$ (i.e. $P \parallel [001]$) and $k \parallel [1\bar{1}0]$ configurations are taken for the bulk Cu_2OSeO_3, and thus, the relation $(P \times M) \parallel k$ is satisfied. (**b**) Experimentally obtained microwave absorption spectra in the skyrmion lattice state at $H = 300$ Oe. (**c**) The difference in absorption spectra between $\pm k$. When the H direction is reversed, the sign of non-reciprocity becomes opposite (Reproduced from; (**a**) Ref. [26], (**b**) and (**c**) Ref. [27])

selection rule, we can identify this mode as being the breathing mode of skyrmions. Later, experimental observation of similar skyrmion resonance modes has also been reported for MnSi and $Fe_{1-x}Co_xSi$ [21].

Because skyrmions in insulators are strongly coupled with a local electric charge distribution (especially an electric dipole), the resonant oscillation of ME skyrmions can be excited not only by H^ω but also by E^ω. Such a magnon excitation active to E^ω is called an electromagnon and has been identified for several magnetic materials with strong ME couplings [22, 23]. When a resonant absorption mode is both magnetic- and electric-dipole active, the interference of these excitations leads to a unique optical response known as directional dichroism. This is a type of one-way window effect, which means that the reversal of the light-propagation direction k provides different absorption spectra. In general, directional dichroism can be considered to be an extension of the linear ME effect into the dynamical regime. By considering the relation $\mathbf{H} \propto \mathbf{k} \times \mathbf{E}$ for electromagnetic waves, the electric susceptibility $\epsilon_{zz} = \partial P_z / \partial E_z$ contains the term linear to $k_z \alpha_{xy}$. This suggests that the absorption spectrum depends on the sign of k ($\parallel z$) when an off-diagonal component of the linear ME coefficient α_{xy} is non-zero. From the viewpoint of symmetry, this condition is satisfied when the relation $(P \times M) \parallel k$ holds [24, 25]. Because $H \parallel [110]$ induces $P \parallel [001]$ for Cu_2OSeO_3, the emergence of directional dichroism can be expected for light propagating along $k \parallel [1\bar{1}0]$ (Fig. 4.5a) [26, 27]. In Fig. 4.5c, the difference in the absorption spectra between $\pm k$ measured for the skyrmion lattice state is indicated. The skyrmion resonance mode shows clear non-

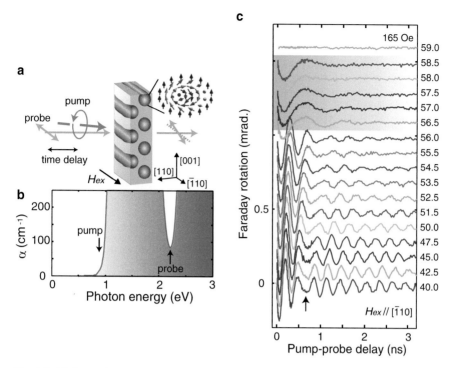

Fig. 4.6 (a) Setup of the optical experiment for detecting the transient Faraday rotation of probe light induced by the inverse Faraday effect of circularly polarised pump light. (b) Absorption coefficient of the Cu_2OSeO_3 sample with the pump and probe photon energies indicated by *arrows*. (c) Temperature evolution of collective spin dynamics excited by right circularly polarised pump pulse. The skyrmion lattice phase is indicated by a *shadowed box* (Reproduced from Ref. [28])

reciprocity up to 3 %, and its sign is confirmed to be reversed on the sign reversal of *H*. The successful observation of directional dichroism inversely proves that skyrmion resonance can be excited by an oscillating electric field, which suggests that the ultra-fast electric manipulation of skyrmions is possible.

Recently, the dynamics of skyrmions has been detected not only in the frequency domain but also in the time domain using the pump–probe technique [28]. Here, a circularly polarised light pulse induced local magnetisation through the inverse Faraday effect, and the ensuing time development of magnetisation was detected by measuring the Faraday rotation of linearly polarised probe light (Fig. 4.6a). The time dependence of the Faraday rotation angle measured at 165 Oe with various temperatures for Cu_2OSeO_3 is plotted in Fig. 4.6c. A clear magnetisation oscillation is always observed below $T_c \sim 59$ K, and the skyrmion lattice state (58.5 K $> T > 56.5$ K) is characterised by a longer magnetisation oscillation period than the helical magnetic state below 56 K. The corresponding oscillation frequency for each magnetic state is consistent with that obtained from previous frequency domain measurements, which proves the successful detection of skyrmion dynamics as a function of time.

References

1. P. Curie, J. Phys. **3**, 393 (1894)
2. M. Fiebig, J. Phys. D Appl. Phys. **38**, R123 (2005)
3. Y. Tokura, S. Seki, N. Nagaosa, Rep. Prog. Phys. **77**, 076501 (2014)
4. T. Kimura, T. Goto, H. Shintani, K. Ishizaka, T. Arima, Y. Tokura, Nature **426**, 55 (2003)
5. M. Kenzelmann, A.B. Harris, S. Jonas, C. Broholm, J. Schefer, S.B. Kim, C.L. Zhang, S.-W. Cheong, O.P. Vajk, J.W. Lynn, Phys. Rev. Lett. **95**, 087206 (2005)
6. H. Katsura, N. Nagaosa, A.V. Balatsky, Phys. Rev. Lett. **95**, 057205 (2005)
7. M. Mostovoy, Phys. Rev. Lett. **96**, 067601 (2006)
8. N. Abe, K. Taniguchi, S.Ohtani, T. Takenobu, Y. Iwasa, T. Arima, Phys. Rev. Lett. **99**, 227206 (2007)
9. Y. Tokunaga, Y. Taguchi, T. Arima, Y. Tokura, Nat. Phys. **8**, 838 (2012)
10. J.G. Bos, C.V. Colin, T.T.M. Palstra, Phys. Rev. B **78**, 094416 (2008)
11. S. Seki, X.Z. Yu, S. Ishiwata, Y. Tokura, Science **336**, 198 (2012)
12. S. Seki, S. Ishiwata, Y. Tokura, Phys. Rev. B **86**, 060403(R) (2012)
13. C. Jia, S. Onoda, N. Nagaosa, J.H. Han, Phys. Rev. B **74**, 224444 (2006)
14. C. Jia, S. Onoda, N. Nagaosa, J.H. Han, Phys. Rev. B **76**, 144424 (2007)
15. J.H. Yang, Z.L. Li, X.Z. Lu, M.-H. Whangbo, S.-H. Wei, X.G. Gong, H.J. Xiang, Phys. Rev. Lett. **109**, 107203 (2012)
16. J.S. White, I. Levatić, A.A. Omrani, N. Egetenmeyer, K. Prsa, I. Zivkovic, J.L. Gavilano, J. Kohlbrecher, M. Bartkowiak, H. Berger, H.M. Ronnow, J. Phys. Condens. Matter **24**, 432201 (2012)
17. J.S. White, P. Prsa, P. Huang, A.A. Omrani, I. Zivkovic, M. Bartkowiak, H. Berger, A. Magrez, J.L. Gavilano, G. Nagy, J. Zang, H.M. Ronnow, Phys. Rev. Lett. **113**, 107203 (2014)
18. M. Mochizuki, Phys. Rev. Lett. **108**, 017601 (2012)
19. O. Petrova, O. Tchernyshyov, Phys. Rev. B **84**, 214433 (2011)
20. Y. Onose, Y. Okamura, S. Seki, S. Ishiwata, Y. Tokura, Phys. Rev. Lett. **109**, 037603 (2012)
21. T. Schwarze, J. Waizner, M. Garst, A. Bauer, I. Stasinopoulos, H. Berger, C. Pfleiderer, D. Grundler, Nat. Mater. **14**, 478 (2015)
22. A. Pimenov, A.A. Mukhin, V.Y. Ivanov, V.D. Travkin, A.M. Balbashov, A. Loidl, Nat. Phys. **2**, 97 (2006)
23. Y. Takahashi, R. Shimano, Y. Kaneko, H. Murakawa, Y. Tokura, Nat. Phys. **8**, 121 (2011)
24. G.L.J.A. Rikken, C. Strohm, P. Wyder, Phys. Rev. Lett. **89**, 133005 (2002)
25. J.H. Jung, M. Matsubara, T. Arima, J.P. He, Y. Kaneko, Y. Tokura, Phys. Rev. Lett. **93**, 037403 (2004)
26. M. Mochizuki, S. Seki, Phys. Rev. B **87**, 134403 (2013)
27. Y. Okamura, F. Kagawa, M. Mochizuki, M. Kubota, S. Seki, S. Ishiwata, M. Kawasaki, Y. Onose, Y. Tokura, Nat. Comm. **4**, 2391 (2013)
28. N. Ogawa, S. Seki, Y. Tokura, Sci. Rep. **5**, 9552 (2015)

Chapter 5
Summary and Perspective

Herein, we will describe several future directions for the study of magnetic skyrmions.

One important issue is the complete understanding of the relation between material parameters (the underlying crystallographic lattice symmetry as well as relative magnitudes of various magnetic interactions such as symmetric exchange interaction, Dzyaloshinskii–Moriya interaction, dipole–dipole interactions and magnetic anisotropy) and the skyrmion-ordering pattern. Despite several promising material design guidelines introduced in Chaps. 1 and 2, the number of material systems reported to host skyrmion spin texture has been very limited. Theories predict that skyrmions can take on various internal spin textures (including vortex, anti-vortex, radiative or hedgehog-like) and ordering forms (including lattice forms with hexagonal, tetragonal and cubic symmetry, isolated particle form and "liquid"-like form) with very different temperature–magnetic field phase diagrams [1–4]. The realisation of smaller skyrmions at room temperature under zero magnetic field would be very essential for future applications of skyrmions to ultra-high-density magnetic storage devices. From a more fundamental viewpoint, it has recently been proposed that nucleation and annihilation of magnetic skyrmions, as well as the associated change in topological number, is controlled by singular magnetic point defects, which can be viewed as quantised emergent magnetic monopoles and anti-monopoles [5]. The concept of a magnetic monopole has been employed to describe several other spin-related issues such as localised excitations in a spin-ice system [6] or an anomalous Hall effect originating from a momentum-space emergent magnetic field [7]. Combined with recent experimental efforts to identify magnetic monopoles [8], a further investigation of their dynamics and responses against external fields as well as the utilisation and enhancement of associated emergent electromagnetic fields will be attractive.

The development of methods to create, annihilate, drive and identify individual skyrmions with ultra-fast speed and minimal energy consumption is another important issue. The creation and annihilation of a single skyrmion has been reported

© Springer International Publishing Switzerland 2016
S. Seki, M. Mochizuki, *Skyrmions in Magnetic Materials*, SpringerBriefs in Physics, DOI 10.1007/978-3-319-24651-2_5

using irradiation by a circularly polarised laser pulse ($\sim$150 fs) [9] as well as using spin-polarised current injection through STM tips ($\sim$1 s) [10], although the latter approach requires a further enhancement of its operational speed. As discussed in Sect. 3.4, spin-polarised electric current in conductive materials can be also used to drive the translational motion of skyrmions, with a threshold current density five orders of magnitude smaller than that of a conventional ferromagnetic domain wall. Although this feature alone will significantly contribute to the reduction of energy consumption, the further suppression of Joule heat loss will be possible through the employment of other external stimuli such as electric fields or spin waves (magnons) in insulators. Note that both spin waves in insulators and spin-polarised electric currents in conducting materials are considered to be a flow of spin angular momentum (i.e. "spin current") [11] and can interact with skyrmions in an analogous manner. In general, spin waves can be excited by various methods such as magnetic resonance [11], thermal gradient [12] and light irradiation[14]. Theoretical studies suggest that skyrmions interacting with magnons propagating along the **k** direction are driven along the −**k** direction and are also deflected along the transverse direction with the skew Hall angle strongly dependent on the ratio of skyrmion size to magnon wavelength [15–17]. These phenomena can be naturally explained by considering the momentum exchange process between skyrmions and magnons under the total momentum conservation law. Recently, the ratchet rotational motion of a skyrmion crystal has been reported for metallic MnSi and insulating Cu_2OSeO_3 under a concentric thermal gradient, which can be understood within the framework of magnon-driven skyrmion Hall effects[13]. The direct experimental observation of translational skyrmion motion under magnon flow or electric field in insulators is yet to be achieved.

Several device structures have been proposed to employ skyrmions in magnetic information storage [18]. They commonly consider the presence or absence of skyrmions at a specific position to be the non-volatile information of a 0/1 bit. One notable prototype was 'magnetic bubble memory', which utilised magnetic bubbles (the rod-shaped ferromagnetic domain as discussed in Sect. 2.2) as an information carrier and was commercially available through the 1970s–1980s. The bubble memory consisted of the interconnection of small bubble tracks comprising a closed loop of guide pieces, reading and writing elements and a driving electromagnet [19]. The magnetic field generated by the electromagnet caused the translational motion of a bubble sequence along the guide. Bubbles were written by the pulse of a local electric current loop (and associated magnetic field) and read magnetoresistively. Because a magnetic skyrmion can be considered to be a type of magnetic bubble, it is compatible with existing bubble devices in terms of structure but with a vastly miniaturised scale and thus a dramatic enhancement of information density. Recently, Parkin et al. proposed a similar but more sophisticated form of magnetic storage called 'racetrack memory' [20]. The element of racetrack memory is made up of a closed loop of a one-dimensional ferromagnetic wire and a pair of reading/writing elements. Here, information is stored as the ↑/↓ direction of a local magnetic moment at that specific position. The application of electric current in the ferromagnetic wire causes a translational shift of information sequence (corresponding to the movement of the ferromagnetic domain wall pattern) through spin-transfer torque. The reading of a local magnetisation direction is possible with a tunnel magnetoresistance (TMR) device, and the writing is achieved by the local injection of spin-polarised current. This simple structure needs only one TMR device per ~1,000 bits, which is regarded as a strong advantage in comparison with an existing MRAM device that requires a TMR device for every bit. By replacing the ferromagnetic domain wall with skyrmions, the critical current density and accompanying energy consumption necessary to drive racetrack memory will be further reduced [21]. Several theoretical investigations support the concept behind such a skyrmion-based magnetic storage device [18, 22], and corresponding experimental demonstrations of the reading, writing and driving of skyrmions confined in nano-structured circuits are in high demand. The development of skyrmion-based information-processing devices is another future issue.

References

1. A. Bogdanov, A. Hubert, J. Magn. Magn. Mat. **138**, 255 (1994)
2. A.B. Butenko, A.A. Leonov, A.N. Bogdanov, U.K. Rößler, J. Phys. Conf. Ser. **200**, 042012 (2009)
3. S.D. Yi, S. Onoda, N. Nagaosa, J.H. Han, Phys. Rev. B **80**, 054416 (2009)
4. J.-H. Park, J.H. Han, Phys. Rev. B **83**, 184406 (2011)
5. P. Milde, D. Köhler, J. Seidel, L.M. Eng, A. Bauer, A. Chacon, J. Kindervater, S. Mühlbauer, C. Pfleiderer, S. Buhrandt, C. Schüte, A. Rosch, Science **340**, 1076 (2013)
6. C. Castelnovo, R. Moessner, S.L. Sondhi, Nature **451**, 42 (2008)
7. Z. Fang, N. Nagaosa, K.S. Takahashi, A. Asamitsu, R. Mathieu, T. Ogasawara, H. Yamada, M. Kawasaki, Y. Tokura, K. Terakura, Science **302**, 92 (2003)
8. M.W. Ray, E. Ruokokoski, S. Kandel, M. Möttön'en, D.S. Hall, Nature **505**, 657 (2014)
9. M. Finazzi, M. Savoini, A.R. Khorsand, A. Tsukamoto, A. Itoh, L. Duúo, A. Kirilyuk, Th. Rasing, M. Ezawa, Phys. Rev. Lett. **110**, 177205 (2013)
10. N. Romming, C. Hanneken, M. Menzel, J.E. Bickel, B. Wolter, K. von Bergmann, A. Kubetzka, R. Wiesendanger, Science **341**, 636 (2013)
11. Y. Kajiwara, K. Harii, S. Takahashi, J. Ohe, K. Uchida, M. Mizuguchi, H. Umezawa, H. Kawai, K. Ando, K. Takanashi, S. Maekawa, E. Saitoh, Nature **464**, 262 (2010)
12. K. Uchida, J. Xiao, H. Adachi, J. Ohe, S. Takahashi, J. Ieda, T. Ota, Y. Kajiwara, H. Umezawa, H. Kawai, G.E.W. Bauer, S. Maekawa, E. Saitoh, Nat. Mater. **9**, 894 (2010)
13. M. Mochizuki, X.Z. Yu, S. Seki, N. Kanazawa, W. Koshibae, J. Zang, M. Mostovoy, Y. Tokura, N. Nagaosa, Nat. Mater. **13**, 241 (2014)
14. T. Satoh, Y. Terui, R. Moriya, B.A. Ivanov, K. Ando, E. Saitoh, T. Shimura, K. Kuroda, Nat. Photon. **9**, 662 (2012)
15. J. Iwasaki, A.J. Beekman, N. Nagaosa, Phys. Rev. B **89**, 064412 (2014)
16. L. Kong, J. Zang, Phys. Rev. Lett. **111**, 067203 (2013)
17. S. Lin, C.D. Batista, C. Reichhardt, A. Saxena, Phys. Rev. Lett. **112**, 187203 (2014)
18. W. Koshibae, Y. Kaneko, J. Iwasaki, M. Kawasaki, Y. Tokura, N. Nagaosa, Jpn. J. Appl. Phys.**54**, 053001 (2015)
19. A. Hubert, R. Schäfer, *Magnetic Domains* (Springer, Berlin/New York, 1998), pp. 588–593.
20. S.S.P. Parkin, M. Hayashi, L. Thomas, Science **320**, 190 (2008)
21. A. Fert, V. Cros, J. Sampaio, Nat. Nanotech. **8**, 152 (2013)
22. J. Sampaio, V. Cros, S. Rohart, A. Thiaville, A. Fert, Nat. Nanotech. **8**, 839 (2013)